MERCURY IN RETROGRADE

Also by Rachel Stuart-Haas

The House Your Stars Built

MERCURY IN RETROGRADE

And Other Ways the Stars Can Teach You to Live Your Truth, Find Your Power, and Hear the Call of the Universe

RACHEL STUART-HAAS

TILLER PRESS
NEW YORK LONDON TORONTO SYDNEY NEW DELHI

TILLER PRESS

An Imprint of Simon & Schuster, Inc.
1230 Avenue of the Americas
New York, NY 10020

First Tiller Press hardcover edition June 2021

Interior design by Matthew Ryan
Illustrations by Shutterstock

Manufactured in the United States of America

1 3 5 7 9 10 8 6 4 2

Library of Congress Cataloging-in-Publication Data

Names: Stuart-Haas, Rachel, author. Title: Mercury in retrograde : and other ways the stars can teach you to live your truth, find your power, and hear the call of the universe / Rachel Stuart-Haas.
Description: New York : Tiller Press, [2021] |
Identifiers: LCCN 2020054921 (print) | LCCN 2020054922 (ebook) |
ISBN 9781982165086 (hardcover) | ISBN 9781982165109 (ebook)
Subjects: LCSH: Mercury (Planet)--Miscellanea. | Astrology.
Classification: LCC BF1724.2.M45 S78 2021 (print) |
LCC BF1724.2.M45 (ebook) | DDC 133.5/33–dc23
LC record available at https://lccn.loc.gov/2020054921
LC ebook record available at https://lccn.loc.gov/2020054922

ISBN 978-1-9821-6508-6
ISBN 978-1-9821-6510-9 (ebook)

CONTENTS

THIS CHART BELONGS TO:

DATE OF BIRTH:

PLACE OF BIRTH:

TIME OF BIRTH:

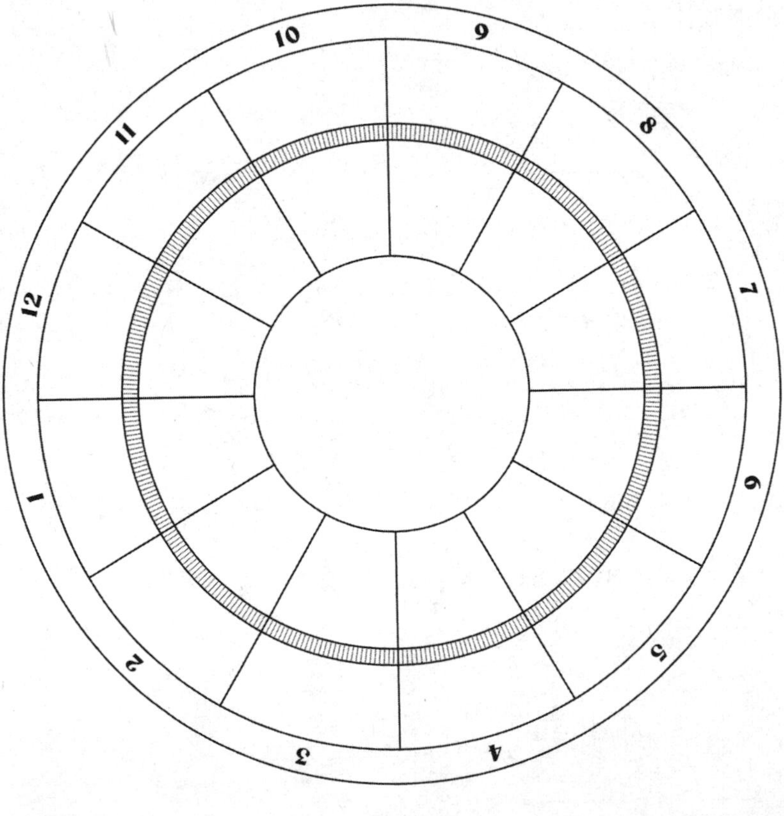

SUN IN: ____________________

MOON IN: ____________________

RISING SIGN: ____________________

PLANETS

MERCURY IN: ____________________

VENUS IN: ____________________

MARS IN: ____________________

JUPITER IN: ____________________

SATURN IN: ____________________

URANUS IN: ____________________

NEPTUNE IN: ____________________

PLUTO IN: ____________________

CHIRON IN: ____________________

CERES IN: ____________________

PALLAS IN: ____________________

JUNO IN: ____________________

VESTA IN: ____________________

HOUSES

1ST HOUSE IN: ____________________

2ND HOUSE IN: ____________________

3RD HOUSE IN: ____________________

4TH HOUSE IN: ____________________

5TH HOUSE IN: ____________________

6TH HOUSE IN: ____________________

7TH HOUSE IN: ____________________

8TH HOUSE IN: ____________________

9TH HOUSE IN: ____________________

10TH HOUSE IN: ____________________

11TH HOUSE IN: ____________________

12TH HOUSE IN: ____________________

NODES

NORTH NODE IN: ____________________

SOUTH NODE IN: ____________________

MOON PHASE

INTRODUCTION

Want to know a dirty little secret? Astrology works. The only issue is that nobody knows how. Perhaps one day very soon, scientists will discover what astrologers have suspected for thousands of years, that the planets have a strong correlation with our daily activities. Astrology is not a religion nor is it any sort of witchcraft. It's the study of the planets, stars, seasons, asteroids, and comets. The planets are affecting you right now. Whether you are aware of it or not.

Sure, we all know folks get a little wild during a full moon. Ocean tides swell and animals scurry about more than usual. It's even crucial for the migration of many birds. One Swiss sleep study found that during the full moon, electroencephalogram (EEG) delta activity during NREM sleep, an indicator of deep sleep, decreased by 30 percent, time to fall asleep increased by five minutes, and EEG-assessed total sleep duration was reduced by twenty minutes.[1] These changes were associated with a decrease in overall sleep quality and reduced melatonin levels. And—get this—those changes happened even when folks didn't know anything about the lunar phase.

And all of this activity can be traced back to our teeny, tiny, itty bitty Moon. So, just imagine how the other planets—some *11.2 times larger* than the diameter of Earth—are affecting our lives.

Is it any wonder that when people are going through a "Saturn Transit" (when Saturn in the sky forms angles or "aspects" to planets in your own natal chart) they feel more realistic, ready for work, and a little bit

1 Katie Wright, "Can a Full Moon Really Affect Your Sleep?" BBC News (May 1, 2018), https://www.bbc.com/news/uk-43960958.

blah. What about a "Venus Transit"? Would you be surprised to learn that when the Planet of Love comes through your chart, people are usually sexier and more sensual, and they love to dress up a bit more? How can we not feel the effects of all these heavenly bodies swirling around in the sky? We are all made of stardust after all.

> **WHAT IS A TRANSIT?** A moving planet is a transiting planet. At every moment, those planets are moving and affecting you, me, and everyone else on Earth. We'll discuss them in more depth in Chapter Two.

Then there's the infamous "Mercury in Retrograde." Retrograde means that a planet appears to reverse direction here on Earth. This optical illusion happens for a few weeks, three to four times a year, and since Mercury is the planet of communication, it's often considered a prime time for conflict and miscommunications.

It's amazing how the placement of the planets at the time of someone's birth can predict their personality, who they're attracted to, even what fulfills their soul. But because we spend so much time talking about our Sun sign, or other elements of our birth charts, we tend to forget that astrology is actually all about free will. Our destinies aren't predetermined. You don't have to jump when the planets say so; no celestial body can force you to think or act in a particular way. They can, though, nudge us in the right direction. And we astrologers would absolutely advise you to listen to the energies of the Universe when it comes to the best time to date, find a job, or ask for a raise. Think of it this way: Imagine there is an island just off a beach. You can reach it with dry feet during low tide, or wet feet in high tide. Nothing is stopping you either way, but if you wait for low tide, the journey will be much easier. The energies and auspices of the Universe work the same way, and if you know how to read the stars you can align your actions with moments when the Universe works for you, rather than against you.

People have been looking toward the sky since the dawn of time. Astrology is an ancient tool that can help identify your highest potential, greatest needs, and even your biggest challenges. Early evidence shows lunar cycle drawings on caves dating as far back as 25,000 years ago. Babylonians and Egyptians have used the powers of astrology; so have kings and queens, presidents, and even billionaires. And you can, too! You don't have to be psychic to learn astrology. It's more like learning a new language—the language of the stars. And the best way to do it is to practice on your friends and family, and often.

Horoscope means *hour watcher* in Greek and is another name for your astrological chart (aka Zodiac chart). When you consult an astrologer to do your horoscope, they examine the Zodiac at your time of birth (and sometimes right now), to ascertain the meaning of each planet, the house it's in, and the sign it's in, to map the challenges or blessings you'll experience in this lifetime. In this book, I'll be sharing tips and tricks to do a foundational reading of your own natal chart. If you're like most people, you don't view life as a series of random, meaningless events, but it can be hard to connect dots, see patterns, and anticipate what to expect. Horoscopes are like having a therapy session with the Universe, allowing us to unplug, connect, and soul search—something we all need in these tech-driven times.

This book will help you understand the basics of astrology and do a deeper dive with the wonderful (and sometimes not so wonderful) events that our Universe can dish out. By the time you are finished, you'll no longer look at an Eclipse or a Retrograde with fear and uncertainty—you'll know these are times for growth, reassessment, and personal evolution, just like when a caterpillar morphs or shifts into a chrysalis. It can help you make major decisions, plan events, and prepare for any upcoming storms when need be. It will absolutely change your life.

If you haven't already, go online and pull up your natal chart. You'll need to know your birthplace and time of birth for the highest degree of accuracy. There are many free sites that provide this information; search "free natal chart" (just make sure to check reviews!) or you can always ask a professional astrologer in your area to make your chart.

Go on . . . I'll wait.

Okay. Got that natal chart? Fabulous!

Let's roll up our sleeves and jump in on what the stars say about you!

1

COVERING THE BASICS

Are you a casual horoscope reader looking to advance your knowledge? Or are you more a bit further down the road of your astrological journey—you know your Sun sign *and* the importance of your Moon and rising? To make sure we're all up to speed, let's go over a few of the basics. We'll be covering planets, signs, elements, and houses here. This may be brand-new information or it could feel a little repetitive, so please proceed accordingly.

THE PLANETS

Okay, we all know that both the Sun and Moon are technically not planets, but astrology doesn't view it that way. To be fair, the Babylonians didn't understand that the Sun is actually a star. They just knew we lived our day by it, and by night—the Moon. Also, I'm sure we are all well aware that poor Pluto has been demoted, but he's still a *big* deal in astrology. And very important. Shhhh. I won't tell him, if you don't.

Now, you can find your own planetary placements by pulling up your natal chart. They're all there . . . go take a peek. Some people have them splashed across the natal wheel, while others might have the majority of them tightly packed in a certain area. Where the planets are located will determine what astrological sign they are in.

SUN

The Sun shows us where we need to shine, and it is hands down the most important planet in a person's chart. It is the center of our Solar System after all, and it's just as prominent in astrology. It shows who we are by sign and placement, what we're here to do, and who we're supposed to be. Is your Sun in the Third House of Communication? Perhaps you excel at writing or sales. What about the work-oriented Sixth House? You're a busybody with lots of little projects. Focus on the Sun first. It is the heart of your astrological chart.

MOON

The Moon represents our feelings, emotions, and natural instincts. It shows how we nurture ourselves, and in turn, how we were nurtured, too. Your mother is represented by the Moon as is your feminine side. Feeling low and gloomy? Focus on that Moon. She'll show you what you *need*. Is your Moon tucked away in the Twelfth House? You need a lot of time alone to recharge. Maybe your Moon is in your Fifth House of creativity. Art can always make you feel better.

MERCURY

Mercury signifies how we think, speak, and talk. It represents how we process information, too. Is it slow and steady in the sign of Taurus? Or sweet and loving in the sign of Cancer? If you have Mercury in the friend house, you gravitate toward chatty folks. Basically, your friends talk. A lot. If it sits in the career house, you talk about work. A lot. Mercury also carries our intellect and shows how we problem-solve, too.

♀

VENUS

Venus displays how we love and what we find attractive and beautiful. It indicates our ideal type of female, the clothes we gravitate toward, colors we adore, and even the music we prefer. Venus also shows where we are charming, graceful, and charismatic. Or *if* we are charming and graceful at all! A well-placed Venus can indicate a highly sensitive, delicate, and even artistic soul.

♂

MARS

Mars is aggressive and masculine. It displays our energy, our inner warrior, and the kind of men we are attracted to. It also displays how we get angry, how intense our temper can be, or if we're more the passive-aggressive type. Mars denotes how we start something and how fearless we can be, too. It's a pushy, brash little guy. A strong Mars indicates a fiery soul!

♃

JUPITER

The luckiest planet there is, Jupiter reveals where we are fortunate and where we feel optimistic in our life. There is, however, a catch: Jupiter can also show us where we are lazy. Why work extra hard when Jupiter delivers so effortlessly? Don't fall for it. Just because the planet of opportunity comes with ease, it doesn't mean we can lounge the day away. Don't stop pursuing your goals, which brings us to . . .

SATURN

Saturn signifies our work ethic and where we really push ourselves. It's ambitious, sure, but really Saturn flaunts our greatest fears and our biggest insecurities. That's why we work so hard in the area that Saturn operates. We're scared to death of failing there, and crazy cautious, too. The smart ones are up for the challenge and utilize the energy of Saturn for growth. The not so wise of us stay away, never giving themselves a chance to face their fears and rise to success.

URANUS

Uranus illustrates where we are original, free-spirited, or how we do things differently than everyone else. Uranus in our chart is where we shun tradition and march to the beat of our own drums. It's where we are rebellious. A strong Uranus can indicate inventors, geniuses, and crazy people, too. Astrologers have a strong-placed Uranus as well.

NEPTUNE

Neptune reveals our dreamy, creative side. It shows us how sensitive and artistic someone is. Or how delusional and drunk they can get. Neptune really doesn't care which way you use it, and sometimes we vacillate between both saint *and* sinner here. The planet of illusions can also hint at where we see things in an unrealistic way. And it's delightfully romantic, too!

PLUTO

Wherever you find the Planet of the Underworld is where you are powerful. This is also the area of your life that you see in strictly black-and-white terms. No gray area here. Where Pluto appears signifies where we are intense, profound, and extreme. It also displays our transformative abilities. Pluto represents death and rebirth, but don't get spooked by that. Remember, it's where we can rise up from the ashes like a phoenix!

THE SIGNS OF THE ZODIAC

The Astrology Wheel is 360 degrees around with each of the twelve signs taking up 30 degrees each. The signs show *the style* in which planets function. Is your Mercury in Cancer? Your style of conversation is sweet, nurturing, and sometimes a little whiny. What if your Mars is in Scorpio? Your style of fighting is intense, sneaky, and manipulative.

Let's not view these as basic Sun sign astrology, which, quite frankly, is horrendously boring and mundane. The signs give depth and character to each planet and their behavior. For example, if you are a Capricorn Sun, don't only get stuck on the definition of a Capricorn. You are much more than that! You may have a fun-loving Sagittarius Moon and a futuristic Aquarius Mercury, too! This makes you interesting and complex. It is so much richer when you can view all the planets in their own signs as a whole. It may be often contradictory, but aren't we all? For now, however, we start more simply. Here are definitions of the twelve signs.

♈

ARIES

"I am"

Aries are the first sign of the Zodiac and of course they must be the pushiest! Aries is ruled by the planet Mars, named for the Roman god of war, and it's just as feisty and brave as the red planet suggests. Aries is also associated with the Ram; these types are strong and stubborn and ready to fight. And fight they do! Especially the younger Aries. Older, more evolved Rams learn to pick their battles, but don't be fooled: *all* Aries enjoy a good challenge. They are born initiators, so sports, combat, and debate are all things Aries can excel in. Anything that hasn't been done before appeals to them, too, because they get to be *first*. That's the best honor that there is! And all Aries secretly want to be *the best* at everything they do. They're highly independent and insanely original, but patience is in short supply here. Real short. That's why their innovative ideas and pioneering spirit can only go so far. If anything gets stale or stagnant, they are already on to the next project. Oh, and did I mention they are impatient?

Aries lives in the here and now. You won't find a typical Aries lingering in the past or fretting too much about the future. And it's true that an Aries can play with your feelings the same way. What? You're still mad about that thing last week? Get over it! It's not that they are cruel or uncaring. Oh no, they deeply want to be liked, and let's face it, they love to show off for you, too! But, like the childlike wonder that they so admirably possess, they can sometimes think only of themselves. But we always forgive them in the end. After all, Aries is far too much fun to have around!

Aries is a sign that jumps into action and the keyword for Aries is "I am." Wherever you have Aries in your chart is where you are brave and where you take initiative. It's also where you need to be the best and can be highly successful.

♉

TAURUS

"I have"

Associated with the planet Venus, Taurus is hands down the most sensual and tactile sign of the Zodiac. Stubborn, too. Here are the creatures that linger over long dinners, taking in all the sights, sounds, and smells of the evening. The clinking of the wineglasses, the deeply rich aromas of the food, the flickering lights of the candles. Truly, I can think of no other sign more fun to dine with! They love to touch and be touched, and so many of them understand the importance of a long, good massage. They are terrible at online shopping, too, since they prefer to feel the fabric, inspecting the quality up close and personal. They loathe cheapness and fast fashion. A true Taurean would rather go without than be faced with knockoffs and poorly made things. They truly appreciate the opulence. All of these high standards would seemingly set them up for a massive amount of debt, but no: Taureans are brilliant with money.

They don't like change, and they are slow to form an opinion, make a decision, or take action. It can be truly agonizing to watch a Taurus start something new (especially for Fire signs). For one thing, they will stay in a situation long past its due date. Mostly because they are "comfortable" and why on earth change that? But once they get going, there is no stopping them. Like a steam train, they have momentum that just won't quit. But when they need a break, watch out. I have known many a Bull whose idea of relaxation is a good full week horizontal on the couch. Preferably with a big bowl of carbs. Other signs find this dull, but not our dear Taurus!

Taurus's keyword is "I have." They are gentle creatures with a strong need for security. Wherever you have Taurus in your chart is where you are stable and grounded.

GEMINI

"I think"

Mercury-ruled Gemini is associated with the symbol of the Twins. They are easily adaptable, and can flow freely in almost any conversation or situation. They are a nervous sign and their keyword is "I think." Geminis are a witty, highly intellectual bunch that can make light of almost any situation. They even feel like a feathery little breeze when they get near you, but don't expect them to stick around for long! They need variety, and loads of it. Boredom is the curse of the Twins. They need plenty to do, to occupy that ever-churning brain of theirs. Books, podcasts, and movies are in constant supply here, feeding their unstoppable need for information.

One of the best traits of Gemini is their wicked sense of humor. Next to Sagittarius, they are one of the funniest signs of the Zodiac. They are also excellent storytellers. Want to wow at your next dinner party? Invite a couple of Geminis and the conversation will never steer dull. The Twins are so well informed on the latest news, gossip, trends. They'll keep you young with their relevant information and their "what's hot now" insight. They're insanely relatable, too. To bask in the glow of a Gemini is an amazing feeling, and they have dozens of friends. If you get even a tad bit monotonous, they will move on quickly, but please don't let it hurt your feelings. They aren't trying to be mean. It's just their brain really can't seem to slow down.

A directionless Gemini is a dangerous thing. Gossip, too, can be a problem for the younger or unevolved Twins. They'll learn to be more sensitive to people with time, though—this is a sign that needs them the most!

CANCER

"I feel"

Cancers get a bad rap. They have a terrible reputation of becoming whiny, sniveling puddles of emotions and clingy to boot! Yikes. Well, truth be told, it's all true. Mostly. They are moody and sensitive but they are also powerful protectors and the most nurturing sign there is. They are the givers of life and if you find yourself on the receiving end of this sign, consider yourself lucky. Oh, sure, Cancers can smother, too. They aren't afraid of feelings and this makes a lot of the signs extra nervous. And they're plain rotten about constantly bringing up the past. They are hopeless romantics who would prefer to live in a more simple and quaint time. The majority are constantly daydreaming.

Cancers tend to jump into action, but they are never direct about it. Cancers are funny little creatures who, like their symbol the Crab, move slightly sideways toward their goals. They are ruled by the ever-changing Moon, and they have a thousand different emotions to match. In fact, like the Moon, their moods change every two and a half days. Their keyword is "I feel" and, oh boy, they really do. They feel *everything*.

They speak in emotions and have strong psychic potential. Cancers seem to just *know* things due to their incredible "gut feelings." But this isn't a confident sign, and when the Crab is even a tiny bit threatened, it will retreat into its shell posthaste. They are convinced that nobody understands them. It's true, nobody really does. Cancers are naturally shy and secretive, making you guess how they feel, then getting all in a tizzy when you don't perfectly grasp their intentions. But in actuality, this makes Cancers brilliantly alluring and highly mysterious.

♌

LEO

"I will"

Leo is blessed with an enviable amount of confidence. It simply oozes from their pores. You know when you're in the presence of a Leo, as you feel as if you're near royalty, and you may as well be! Leo's symbol is the Lion. Majestic, vain, and proud. They are ruled by the Sun, and you'll never feel so warm as the warmth a Leo can give you. They are very affectionate and love to spoil with lavish gifts. The bigger, the better!

The talent these creatures possess is nothing if not extraordinary. The problem is, Leo is well aware, and they'll tell just about everyone they meet all about it. Humble is not in Leo's vocabulary. They're terrible show-offs. But it's so elegantly accomplished, you almost enjoy the dramatics of it. They make wonderful actors and artists, but what's the point if they don't have an audience? They'll do anything for the sound of applause, and they have an uncanny knack for putting a creative flair on *everything*. Let's be honest. Life is more fun with a Leo.

They have a lot of pride, too, and their egos are fragile things. If the spotlight isn't shining on them, they can behave like a terrible toddler. Ignore them and it gets worse. Really it can be like a no-win situation. Fawn over them. Once a Leo feels he or she has your attention, they're deeply loyal. Not to mention generous, too! They have big hearts, big hair, and big style. Leo's keyword is "I will." Wherever you have Leo in your chart is where you are incredibly confident and can show off.

♍

VIRGO

"I analyze"

They seem so sweet and docile, don't they? At first, Virgos seemingly go with the flow. All dressed to perfection and always right on time. But behind that pretty smile of theirs are heaps of nervous energy and more insight than you can shake a tail feather at. You see, they don't miss a thing. It's the blessing and the curse of being a Virgo. They have eyes everywhere and their attentiveness is sometimes overwhelming. For you and for them. It's true they are perfectionists. They see the most minute details and can barely contain themselves to point them out. Some act ashamed of this. With lots of "I'm sorry's" and "excuse me's." They know it's maddening, but to be honest, we *need* Virgos.

After all, who else will point out that the cup in the cupboard isn't facing the same way as the other mugs? How else will we know that the zipper on our skirt isn't *exactly* aligned with our spine the way it should be? Thank God for Virgos! Sure, they are critical, but here's the twist: they're first and foremost critical of themselves. They have exceedingly high expectations of themselves and of you. Virgos truly help us become better people.

They're natural healers and teachers, and they have a deep, soul-fulfilling need to be of use. They are always happy to help. Even when you don't want them to. But trust me . . . you do. They analyze everything and are very, very smart. Like Gemini, Virgo is associated with the planet Mercury. Their symbol is the Virgin, invoking purity and perfection, but don't let that fool you: they have a seductive side just like the rest of us.

♎

LIBRA

"I balance"

Libra's symbol is the Scale, always striving for balance with perfect fairness and equality. An impossible feat, but try they must! Aligned with the planet Venus, they are well-mannered, beautiful creatures who loathe conflict more than anything in this world. Their keyword is "I balance" but it could just as easily be "peace at any price": conflicts of any kind drive them batty.

Libras naturally see both sides of a situation. Don't even try to get them on *your* side. They'll eventually feel badly for the other guy, too. It's all part of the balance! This makes them wonderful counselors, diplomats, and mediators. They seek harmony and equality above everything. You'll see it in the way they dress, decorate their homes, and how they conduct their business. It's all very fair and perfectly symmetrical.

It's a good thing so many of them are so sweet and exquisitely beautiful, too, for Librans thrive in relationships. Whether it's romantic or corporate, Librans *need* a partner. Being alone is a Libra kryptonite. It throws their scale way off and then they'll have no one to bounce their brilliant ideas off. Maybe they should take up tennis.

Librans are social extroverts. They have to watch out for being less than truthful just because they don't want to hurt someone's feelings. They'll learn one way or another. And one can never stay mad at Libra for too long; they're far too sweet and charming!

♏

SCORPIO

"I create"

Scorpios may be one of the most difficult signs of the Zodiac to understand. That's okay; they barely understand themselves at times. They are ruled by the planet Pluto, god of the Underworld, and their emotions are just as devilish and deep as the god's home suggests.

Their key phrase may be "I create," but let's be clear, Scorpios can destroy, too. Sometimes I think the symbol of a phoenix would far better suit a Scorpio than an actual Scorpion, since these signs are so aligned with both life and death. The less evolved ones are jealous, manipulative, and vindictive. The evolved ones have obliterated their old ways and re-emerged as almost angelic. This is the beauty and the mystery of Scorpios. As Carl Jung wrote, "No tree, it is said, can grow to heaven unless its roots reach down to hell."

Their experiences are met with high highs and low lows. Life is never half-assed with a Scorpio. Perhaps that's why so many of them make excellent psychologists. They aren't afraid to venture into the dark and forbidden, and that includes your psyche.

Scorpios have surprisingly good memories, too. They never forget a compliment or a critique. They won't bring it up like a Cancer would, but you can see it in their eyes. A wounded Scorpio is a scary sight. It's true that Scorpios need to work hard at not being *too* judgmental, but they can't help it, really. They have an incredible stealth system of values. A system that you best not cross.

Scorpios are highly creative and exceedingly passionate creatures. They are the most resourceful sign there is. Gain their trust and you have a loyal friend for life. Wherever you have Scorpio in your chart is where you do things in extremes. These are the aspects of your life where you are powerful and intense, and don't take things lightly.

SAGITTARIUS

"I achieve"

Sagittarians are ruled by the planet Jupiter, the largest planet in our Solar System. And just like Jupiter, Sags do everything *big*. Here is an optimistic sign with big ideals, big excitement, and a big mouth. They see the big picture, too. In conversation and in worldly principles.

Plain and simple, Sagittarians are fun. They'll take the most mundane task and turn it into a celebration. They're great communicators if you can ever get them to shut up. They'll prattle on and on, telling stories from all their many travels, or from the documentary they saw on Netflix last night. A lot of them can be know-it-alls, but the majority of them are so funny, who really cares? Sags are highly intelligent, with such cheerfully crafted insight it's no wonder they are so popular!

The symbol for Sagittarius is an arrow. They shoot it far and wide, which is why they're so famously known as the adventurous bohemians. Viewing themselves as sort of eclectic gypsies, they are the wanderers of the Zodiac. Sags love all forms of mind expansion, too. They are avid learners and make great teachers, but they loathe responsibility and being tied down. Freedom is paramount here. Actually, it's not so much that they are irresponsible per se; they just strongly dislike being told what to do. It ruins their whole free-spirited vibe and brings them down. A sorry sight to see indeed. Thank goodness they never stay down for long! A healthy Sag always bounces right back up, yet another reason they're so inspirational!

At their worst, Sags say more than they should. Don't tell them too many secrets. They'll use it for a funny story and wonder later why you are seething at them. But truth and honesty are ever so imperative to these Centaurs. They'll tell it like it is, even if it's a tiny bit hurtful. Wherever you have Sag in your chart is where you're honest and optimistic, too. It's also where you can take a gamble. Their key phrase is "I achieve."

♑

CAPRICORN

"I use"

The old souls of the Zodiac are Capricorns. They're the hardest workers, too, but please don't tell the Virgos! Capricorns are ambitious and opportunistic, unafraid to jump in and get their hands dirty. They know what it takes to do a job well, and they definitely aren't afraid of a challenge. Capricorns are associated with Saturn. The planet of handwork and diligence. Its phrase is "I use."

Capricorns are a serious bunch. Like the Goat that is their symbol, they'll climb impossible mountains with effortless concentration. Try as you might, you can never distract a Capricorn from the goal. And let's be clear, the goal is always "success." That's why so many Capricorns are in fact very, very successful. But it's not sheer luck that got them there. They have to earn their fortune. To quote Ralph Waldo Emerson: "The reward of a thing well done is having done it." The younger ones are wound tight. They have heaps of expectations, usually self-imposed, and take the world very seriously. The older ones have loosened up over time. They age like fine wine and find the joy in life instead of the hardships they've seen before. They can be funny, too! The dry, sarcastic wit of a Capricorn is nothing if not hilarious. They excel at self-deprecating humor. Best to beat you to the punch! They're highly aware of their flaws, and they'll work doubly hard to distract you from it. It's one of Capricorn's biggest weaknesses.

These sweet souls can get depressed, too. The world is a harsh place, and Capricorns are insightful and whip smart. They dwell in reality, and have a hard time with the illusions of fantasy. Help them out with distractions of books and movies. Get them to take breaks, and they'll thank you for it.

AQUARIUS

"I know"

It's no secret that Aquarians are the weirdos of the Zodiac. It's why we love them! Oh sure, they're detached, aloof, and . . . odd. They can come off as uncaring and standoffish, but they are independent and exceedingly original and light-years ahead of you and me. No wonder they seem so easily bored with the rest of us!

Aquarius is linked with the planet Uranus (*and* Saturn before we even discovered Uranus). The key phrase for Aquarius is "I know," which is fitting since they don't run on feelings or instincts. Aquarians have pure electricity running through their veins and they are hooked up to a futuristic world that you and I aren't invited to . . . yet. This is why so many of them are inventors in one way or another. They have new concepts, new frameworks, and new ideas. They break the mold, these Water Bearers.

They are rebellious, too. They loathe authority, and refuse to fall in line with the crowd. Paradoxically, they are in fact amazing team players, even if they stick out like a sore thumb. Aquarians just know it takes a village, so they'll play along with everyone else for the sake of the future. They know you can't change the world with only one person. They're natural advocates and activists, and they'll fight for a cause with gusto. Shattering old structures and ideas, and reinventing new ones in its place.

Aquarians are extraordinarily charismatic and individualistic and make wonderful friends. They usually run in several different crowds, unable to meld with just one sort of social type. They love their freedom and flexibility, so give them that and they'll be forever in your debt.

Wherever you have Aquarius in your chart is where you are independent and futuristic. These are the places to experiment!

PISCES

"I believe"

The last sign of the Zodiac is also the most mystical. Pisces is associated with the planet Neptune and its symbol is that of two fish. This is an immensely sensitive and psychic sign. True, they seem so fragile and soft, but Pisces is anything but. You see, they don't live in the same world as you and I. Pisceans feel most comfortable fully submerged beneath the conscious layer of life. They dwell in dreams and fantasies, and sometimes a good bottle of red wine. They see behind the veil and speak to the strange and unusual world behind it. They are the most ethereal and dreamy of all the signs.

They are natural empaths and they feel others' pain, whether it be human or not. They sympathize and cry when you do. Then they laugh and cheer with you as well. They make great allies and the best of friends. Their secret power lies in their ability to assist. They know it, too, for when a Pisces serves it halts their own suffering. And boy can these fish suffer!

Martyrs and victims make up the less evolved Pisces. Whatever it is, it's never their fault: they were too weak and helpless. They thrive on drama and sulk around eating crappy food, popping pills, and smoking cigarettes. Addictions run high in this sign. When they learn to help the helpless, they give up some of their own self-pity and become useful.

The more evolved of the Fishes run on faith and romance. They believe in a higher power and that everything will work itself out just right. Just as it's supposed too. Karma is their friend and wherever you have Pisces in your chart is where you believe and have faith, too.

ASTROLOGY & THE ELEMENTS

There are four elements at work in astrology—Fire, Earth, Air, and Water—and each of the twelve Zodiac signs is aligned with a particular element:

FIRE: Aries, Leo, and Sagittarius

EARTH: Taurus, Virgo, and Capricorn

AIR: Gemini, Libra, and Aquarius

WATER: Cancer, Scorpio, and Pisces

The elements play a big role in your personal cosmic makeup. For example, you're a Fire sign if you are an Aries, but if you have heaps of other planets (Venus, Mars, Jupiter, etc.) in the Water elements, you will be a more sensitive and empathic Aries. The elements garner a different interpretation and depth to your Astrological Chart.

Some people only have one or two planets in each element; others have many. If you lack any of the elements, and people often do, you'll find yourself trying to overcompensate in that area. Perhaps you also attract people who can "fill in" that element for you. For instance, a chart with very little Air might seek out a few good Geminis and Libras to help them learn the fine art of socializing and conversation. Lacking Fire? It might be hard to start new things, but you can learn a thing or two from an Aries or Leo. Earth? Practicality is not your strong suit. Capricorns and Virgos can help you out. Lacking Water? You are learning to be more empathetic; seek a Cancer.

FIRE

ARIES, LEO & SAGITTARIUS

Fire folks are enthusiastic, confident, and ready for action. They have a "me first" attitude that takes them far in life but is sometimes criticized as . . . ahem, selfish. But that's okay—the vast majority of people with Fire-heavy charts don't really care what you think anyway. They've already moved on to the next thing! This self-centeredness doesn't mean that they're unkind, though. In fact, they are courageous, with incredibly high standards. Before there is fire, there must be a spark, and Fire signs excel at starting new things. Finishing them, though, is an entirely different story.

EARTH

TAURUS, VIRGO & CAPRICORN

Earth signs are grounded, security driven, and practical. Need a job done well? Get an Earth sign on it! They are resourceful, reliable, and hardworking. These are loyal signs, and they deeply need to be useful. They may seem conservative, and let's be honest, some of them are. But here is a group who secretly and sometimes not so secretly harbor a lusty, sensual appetite. They appreciate the finer things in life and love money. It is to them what makes the world go around! They can't stand being uncomfortable and not having nice things. No wonder they seem so successful. They work hard for what they want.

AIR

GEMINI, LIBRA & AQUARIUS

Like little hummingbirds, zipping around from one thing to next, Air signs are social and expressive. They're very bright and naturally intelligent. They love to chatter to you and everyone else they come into contact with. It's almost flirty how inquisitive they are! Logic rules, though, and emotions are something they sometimes are too busy to deal with. They'll rationalize it away, make lists, weigh pros and cons, too. Funny enough, Air signs can go berserk when being forced to make decisions. That's when they call up all their friends to "talk." It's rare to be an introverted Air sign.

WATER

CANCER, SCORPIO & PISCES

Ever need a shoulder to cry on? Someone to listen without judgment or trying to fix your problems? Go find a Water sign! These guys are emotional and highly empathic. But don't let their soulful eyes and sweet caresses fool you. They can be annoyingly moody. Water signs end up focusing so much on others, becoming comforting and supportive to distract you from figuring them out. They excel at keeping their true emotions hidden below the surface. But you can't really blame them. They speak in feelings, and have a hard enough time trying to understand their own. Let them fuss over you! They love to nurture.

CARDINAL, MUTABLE & FIXED

There are three different qualities to the signs. We begin with the Cardinal signs. They are the initiators and the first to react. Aries, Cancer, Libra, and Capricorns are the ones that perform quickly and make up the Cardinal quality.

Mutable signs are more "go with the flow" and flexible. They may seem "whatever" and easy. Gemini, Virgo, Sagittarius, and Pisces are the Mutable, more flow-y signs.

The Fixed signs are stubborn and hard to get going. Taurus, Leo, Scorpio, and Aquarius make up this more stationary (sometimes hard-headed) group.

THE ASTROLOGY WHEEL

There are twelve signs and twelve houses in the Astrology Wheel. Each house represents a certain area of life and is also associated with a sign and a planet. Think of each astrological house like a stage where the actors (planets) get to express themselves. These "stages" each represent a specific environment. The signs that the planets are in become the type of character they play. The houses are simply where the scenes take place.

We start with the First House, which is also the Ascendant, or Rising Sign. The Ascendant (AC) is the first thing people see when they meet you; it's a bit like the mask that we wear. It's not the *real* you, it's your defense mechanism and how you want people, especially those you don't know very well, to see you. The Ascendant denotes the way we look, how we dress, even how we carry ourselves.

The opposite of the Ascendant is the Descendant. The Descendant (DC) begins on the Seventh House and illustrates the kind of people we attract. The line that cuts across the chart that marries both Ascendant and Descendant is actually the horizon line from East to West. The houses below the horizon line are inclined to more personal needs, while the houses above the horizon are affiliated with more worldly affairs. In a nutshell, if you have loads of planets up top, you have a knack for knowing what the world expects of you, and you're happy to play the game. All below? You've created a rich, personal, inner world and may not care less who knows about it. But in all likelihood, you'll have planets in both.

The tippy top of the chart is the Midheaven or MC (Latin for *medium coeli*). It displays our public reputation and how we want to be known in the world. In contrast to the Ascendant, which is more personal and one-on-one, the MC reveals how we function in a group and how we want to be known in our career. Below the MC, we find the IC at the bottom of the chart. The IC (Latin for *imum coeli*) is the beginning of the Fourth House and it reveals a hell of a lot about where we come from. Debbi Kempton-Smith, a brilliant astrologer, taught me that the line between the MC and the IC is more like the spine of our chart. It's just as imperative as the Ascendant.

As you begin to learn about the houses in your own chart, don't fret if you don't find any planets in a particular house. It doesn't mean you are a lost cause in that area. They're called Empty Houses, and you'll probably get by without a lot of extra work in that particular house. You can be just as successful in that area with or without planets there.

Of course, you'll need a computer to find out where your planets are located and which signs you have on the houses in your own chart. But for now, let's go through each house, starting at the First and continuing counterclockwise.

FIRST HOUSE

SELF

The First House is the doorway to the chart and it represents our identity, our body, and our health. This is where your Ascendant is. It shows our first response and our first impressions, plus our views of the world in general. This is your persona house. It's the you that you put on display. Sometimes it's similar to who you really are, sometimes not. Planets in this house are the first responders. Got the Moon here? You respond to things emotionally. Mars? Anger and action move you. The First House is ruled by Aries.

SECOND HOUSE

MONEY

The Second House is also known as the Money House and it shows us how we make our income, but it also reveals what we value. It even reveals how we value ourselves. We'll find security and safety here. Our prized possessions and all our stuff. Some people like to share and others are more greedy. Some folks have to work more for money, and others are just plain lucky. The Second House is ruled by Taurus and is aligned with material things and sometimes food.

THIRD HOUSE

COMMUNICATION

The way we communicate is the basis of the Third House. It's ruled by Gemini and indicates the way we speak and how our minds work. It shows our perception of things and how we process information. It also hints at our relationship with siblings, if any, and our neighbors. It rules short-distance travel, and involves our preferred mode of transportation. Any way you can use your voice is shown in this house.

FOURTH HOUSE

FAMILY

The Fourth House is where the IC is located and it illustrates our family and childhood experiences. It reveals what sort of home we feel comfortable in and even how we like to decorate our abode. It's real estate and land and all things house and home related. The Fourth House is also always one of the parents. Mostly the Father, but astrologers argue about this quite a bit. It really depends on who was the dominant parent. What's really interesting is that this house also hints at how we function in old age.

FIFTH HOUSE
PLEASURE

The Fifth House is fun, babies, art, and romance. It's the house that shows your childlike side and how you play in life. It's love, too. How you give it and how you receive it. If you have children, this is the house of your first-born and how you function as a parent. It also brings in any hobbies you like, sports, and games. Gambling is in here, too. Are you a risk-taker, or do you play it safe? The Fifth House is ruled by Leo.

SIXTH HOUSE
ROUTINE

This is the busybody house and shows the way we work, our schedules, and our routines. Health falls into this house, too, as you can't really be robust and strong without a good routine, right? This house also indicates what types of coworkers we work with and what kind of food we like. This is the House of Service, ruled by Virgo, and I've seen many nurses, yoga instructors, and massage therapists with a strong Sixth House. This house wants to be of use.

SEVENTH HOUSE
RELATIONSHIPS

The Relationship House is ruled by Libra. A lot of folks call this the Marriage House, but truth be told it tells us about every single person you bring into your life. The Seventh House really signifies the sort of people we attract, mostly because we think we are deficient in that area. As if we need to fill a void in our own chart. Wanna know a secret? If you are feeling lonely, focus on your Seventh House by sign and by planets—if any. You won't need anyone else to fill you out.

EIGHTH HOUSE

SHARED ENERGY

This is the Psychology House, ruled by Scorpio, but really it's the home of shared energy. Sex and money fall into this category because that's what they are: shared energy. This house indicates how deep we can go into our psyches, and how sheepish or wild we are in the bedroom. It's all things taboo and this includes death and deep transformation. If you are in a partnership, this house will reveal a thing or two about the other person's financial situation.

NINTH HOUSE

PHILOSOPHY

The Ninth House is ruled by Sagittarius and it is associated with long-distance travel, religion, philosophy, and how people talk to you. This is the house that shows what kind of vacations you like, and which far away lands you want to visit. It indicates your fundamental truths and what you believe in. It's publishing, too. If the Third House is how you speak, the Ninth House shows who's listening, and how they respond. It's college, the law, and even your in-laws. This is the Mind Expansion House.

TENTH HOUSE

WORK & PUBLIC RECOGNITION

The Tenth House is where the Midheaven is, and is ruled by Capricorn. This is our public image, our career, and what we want to be known for. Fame is here and power, too. It's authority figures, from your boss to your mom, and it displays your responsibilities in this world. It rules our ambitions and reveals the direction we are going in our life. Both the Tenth House and the First House are what we put out into the world. It's the most "showy" of all the houses.

ELEVENTH HOUSE

COMMUNITY

This is the house of friends, community, and our goals for the future. It's your peers and the social circles you gravitate to. It also shows how your friends treat you and how you make the world a better place. It's your hopes and your dreams and all the wonderful plans you have for a better tomorrow. This house is ruled by Aquarius and indicates how well you are a team player or not.

TWELFTH HOUSE

THE UNSEEN

The Twelfth House is ruled by dreamy Pisces and it's the house of everything hidden. From you and everybody else. This is the home of the unconscious and where we deposit all those things that we can't possibly deal with today, so we'll think about that tomorrow. Oh, it'll come up sooner or later, and it's often quite healing when it does. There's a great deal of self-awareness when you really look inside the Twelfth House. This house is spiritual, not religious. It's faith and psychic abilities and everything that exists on the other side of the veil.

Your natal chart never changes. But the planets in the sky are constantly in motion. Moving and swirling around up above, they "aspect" your own natal planets but they also make their way through each of your individual houses too. For example, as Mars rotates around your chart, it will light up each one of your houses or "stages," activating and energizing each house.

Polarities in Astrology

Every sign has a polarity or opposition. It's simply what's across the way on that big ol' Astrology Wheel—think of it a little like a color wheel. Leo is opposite Aquarius. Gemini sits opposite Sagittarius. If, say, you're a Virgo, Pisces is your polar opposite. Your opposite sign will bring to the table what you lack. Aries, for instance, is independent. Rely on someone else? Never! But Aries can miss out on the beauty of compromise. Enter its opposition Libra, who thrives in relationships. But even a well-balanced Libra can learn a thing or two from a self-reliant Aries.

You can learn quite a bit from your opposite sign. Here are the polarities:

ARIES — LIBRA

TAURUS — SCORPIO

GEMINI — SAGITTARIUS

CANCER — CAPRICORN

LEO — AQUARIUS

VIRGO — PISCES

2

UNDERSTANDING TRANSITS

When astrologers talk about "transits," it may sound fancy or complicated, but a transit is simply a planetary movement that is happening in the sky at any particular time. Technically, a transit is "the act of passing through" and we use transits to interpret predictions by comparing what's happening in the sky now and comparing it to our own natal chart. When an astrologer looks up your transits, he or she is mapping the planets, wherever they happen to be, into your own natal chart. Then we can see how the planets are "passing through" your very own individual blueprint.

Planetary Transit Timeline

Each planet moves at its own speed and pace around the Astrology Wheel. You'll notice the farther the planet is away from Earth, the longer its orbit will take.

SUN: 1 year around Astrology Wheel
MOON: 28 days around Astrology Wheel
MERCURY: 1 year around Astrology Wheel
VENUS: 1 year around Astrology Wheel
MARS: 2.5 years around Astrology Wheel
JUPITER: 12 years around Astrology Wheel
SATURN: 29.5 years around Astrology Wheel
URANUS: 84 years around Astrology Wheel
NEPTUNE: 165 years around Astrology Wheel
PLUTO: 248 years around Astrology Wheel

From the moment we are born to the age we are now, we've gone through hundreds of transits that help us grow and evolve. A Saturn transit will mature us. A Pluto transit will transform us. A Uranus transit will liberate us, and so on. Think of each transit as a golden opportunity to work on that particular area in your life. They aren't simply to be endured, though sometimes it often feels that way! Some transits occur frequently, while others are a once-in-a-lifetime deal. The ones that happen often are the Sun, Moon, Mercury, Venus, and Mars, which we call the Inner Planets. Jupiter, Saturn, Uranus, Neptune, and Pluto make up the Outer Planets and they occur less frequently, but also amass a much more powerful punch. The shorter the transit, the less time you spend incorporating the energy of that planet in your life. This is in contrast to the Outer Planet transits, which can take months or even years to complete.

SUN

The Sun takes exactly one year to go around the Astrology Wheel. It's like clockwork in that you know that the same time every year the Sun will light up that particular house, and/or planet in your chart. Every March 21, for example, begins the sign of Aries. So, wherever you have Aries in your chart is where that Sun will be every March 21.

The Sun brings energy and awareness to each house every thirty days. It's fun to experiment with this, and you can even plan out your year by charting the Sun's transit. For instance, if you have the sign of Libra in your Third House of Communication, we know that every October is great for writing and getting yourself heard. Play with this and you'll see a pattern emerge. Perhaps every February you notice you shift your career around. Or every June you seem to focus more on your friendships as the Sun passes through your Eleventh House of Groups of People. You can also learn what not to do during the Sun's transit. Don't plan a big event when the Sun is transiting your Fourth House, for example. You're more in the mood to just stay at home.

MOON

The Moon takes about twenty-eight days to go around the Astrology Wheel, changing signs every two and a half days. She moves rather quickly and is all about the *feels*. The Moon indicates our emotions. She activates our moods and reveals a time for reflection. I personally watch the Moon like a hawk, and if you have a Moon-ruled Cancer Sun, Moon, or Ascendant, you should, too. These folks literally change their disposition every two and a half days just like the Moon.

Every time the Moon changes signs she goes through a period called a void. This can last anywhere from a couple of minutes to a couple of days. Basically, a void is determined when the Moon has made her last aspect before shifting signs. Eerily, plans seem to fall through during voids and items purchased don't see their intended use. It is, however, a great time for sleep, meditation, creative pursuits, and yoga.

As the Moon swims through each sign, it sets the atmospheric stage that you and I and everyone else are feeling. For instance, people are more fun during Fire Moons, more serious during Capricorn Moons, and more giving during Pisces Moons. Now see where the signs fall in your own chart. Perhaps you have Scorpio in your Sixth House. Once a month, when the Moon is there, it's your best time to get caught up with organizing and cleaning. You may also kick up your exercise routine a notch. Since it's in Scorpio, you'll be pretty damn intense about it, too.

MERCURY

Mercury is never more than 28 degrees from the Sun. So just like the star of our Solar System, Mercury takes approximately one year to go around the Astrology Wheel. This planet is quick and when he zooms through our chart he brings a chatty, curious, busy energy along for the ride. Mercury is known as the Messenger God, so there's usually some sort of new information or fresh message left in his wake.

Mercury transiting your Moon? (Translation: This means Mercury in the sky is crossing paths with your natal Moon.) Perhaps you get some news from your mother or a female friend. Maybe you'll just want to talk about your emotions, too. Mercury on your Saturn? This can mentally stimulate your own fears and insecurities. While in contrast, transiting Mercury to your Jupiter brings good news and big ideas. As Mercury speeds through the houses, we can expect to talk a lot about the subject that Mercury is in. We chat a lot about money when Mercury is moving through the Second House. And maybe we feel inspired to talk to other people more when Mercury visits our relationship house.

A Mercury transit doesn't last for long, unless we fall into the dreaded Mercury Retrograde, which need not be feared, but we'll get to *that* later.

VENUS

The Planet of Love is never more than 48 degrees away from the Sun, and similar to the orbit of Mercury, it takes approximately one year to go around your chart wheel. Venus brings along a lovely, sweet, and charismatic vibe when she visits. Basically, Venus just feels good. Venus wants us to feel pretty and attractive, loved and cherished. She wants us to eat cake, drink pink champagne, all the while with charm and grace and ease. Her visits are unfortunately all too quick, so grab her lovely energy while you can!

Venus transits your Sun once a year. It's a joyous day where you feel at peace and loved. Venus moving across your Mars is hyper-romantic, a super-sexy day where you look and feel amazing. Venus can help you meet the person of your dreams. Is she moving through your Fifth of romance? Don't sit at home! Someone's waiting for you out there! Same with Venus, your Seventh House. Of course you'll have to note where all the other planets are, especially the Outer Planets. But a good Venus transit can be gold.

MARS

The planet of action moves through the Astrology Wheel about every twenty-two months and stays in each sign approximately every six weeks. So, if you've ever wondered why so many two-year-olds go through the "terrible twos," now you know! They're going through their first Mars Return! Basically, this means Mars orbits back to the exact location it was at the time of birth. It's a feisty, aggravated time and we all go through them every couple of years.

When Mars comes in for a visit he brings out anger, aggression, and vigor. He keeps us busy and ready to jump in the action. He can stir things up, causing friction and fights, but remember: You *always* have free will. Mars is pushy, impatient, and demanding, but you get to decide if you prefer to argue or start something new.

I think of Mars as an activation planet. He'll come in and light up our natal planets and stir up our houses. Mars is that spark that we need to peel ourselves off the couch. Mars simply wants action. It gets us moving and is the most fantastic planet to help us get things accomplished!

JUPITER

The luckiest planet takes about twelve years to go around the Astrology Wheel. This is the transit that brings glorious opportunities and phenomenal people into your life. It'll bring you love, a raise, a new house, and a bigger waistline. It ain't all positive here! Jupiter is the planet of expansion and whatever planet he touches on, or whichever house he is passing through, gets bigger, brighter, and way more pronounced. And yes, that can include our fears, our anger, and our delusions.

Jupiter can help us plan the best times to travel and the best time to go back to school. It'll absolutely always give us a bigger perspective and a lot more confidence, too. The downside here is that sometimes Jupiter feels *so* good and is *so* lucky, we halt our drive and effort. We already have lady luck on our side, so surely we don't need to work, right? Wrong. Don't let Jupiter make you complacent.

Jupiter can also garner us a lot of attention. Watch out for that, and make sure you're doing the kinds of things you'd like to get recognition for. It's almost impossible to sneak around when Jupiter is in your Tenth or First House. Jupiter can be quite exposing.

SATURN

The Taskmaster Planet takes about every twenty-seven to thirty years to go around the Astrology Wheel and this includes the famed "Saturn Return." It's the planet of maturity and growth and there's always some sort of suffering or hardship when we go through any heavy Saturn transit.

In its best form, Saturn helps us evolve and blossom into a responsible soul who takes pleasure in our accomplishments. At its worst, Saturn is depressing, cold, and lonely. A heavy Saturn transit is a burden and at times feels impossible and never-ending, like Sisyphus pushing his giant boulder up the mountain every day. Even so, Saturn transits are crucial for our development. Yes, they are challenging times, but as my friend and astrologer Quan Tracy Cherry once told me, Saturn is like Domino's pizza: You work hard, and it delivers.

URANUS

The Planet of Shock and Surprise comes around every eighty-four years—a once-in-a-lifetime event!—and always delivers a great deal of transition plus a little bit of anxiety, too. Well, after all, change *can* be pretty nerve-racking! Uranus is freedom, liberation, and revolution, and he'll infuse us with those very same sentiments, too. Right before a strong Uranian transit, we often feel bored and restless, as if life has become dull and rather monotonous. But not for long when Uranus is headed our way!

A strong Uranus transit will inspire us to quit our jobs, move across the country, and perhaps take in a new lover. It will leave our friends and family scratching their heads. Let's face it, most of us would never have done anything like *that* before. This is why Uranus is so fun. We become unpredictable, and also insanely impulsive. Our views shift, as do our tastes. We want to be wilder, more free. Here is the downfall of Uranus. He wants us to act fast and act *now*, and that can be a little dangerous. I've known many people fall under the spell of Uranus only to regret it later, so make sure the change you crave is also a smart one.

The best way to utilize this transit is to remain flexible and open for anything. Fasten your seat belts, people; it could be a very bumpy and exciting ride.

NEPTUNE

The dreamy planet of Neptune takes 165 years to go around the Astrology Wheel. It comes with a heavy dose of compassion, empathy, and a whole host of delusions as well. It'll have you picking up the drunks at the bar and sleeping with them, too. You'll want to save everyone. Your friends, your coworkers, and that cheating ex. You can't, of course, and you shouldn't try. Oh, reality is so hard to see when one is enchanted by Neptune.

Escapist habits hit hard when Neptune is in town. You can't seem to see straight, and everything's a blur. You can't remember where you've been or even where you're going. Plain and simple, this transit will leave you confused. Neptune is the planet of dissolve. That could mean your marriage, your career, or even your bank account. It's the fact that we don't seem to pay attention when Neptune is in for a long visit. This is fixable and temporary; thankfully every cloud does have a silver lining.

Neptune can also inspire and turn up our creative juices. It can unlock spiritual pathways and assist us in a higher purpose. You'd better believe intuition is enhanced and psychic abilities grow strong under the influence of Pisces-ruled Neptune.

PLUTO

Pluto takes 248 years to go around the Astrology Wheel, so some people will never experience a major Plutonian aspect. Which is a pity: there probably isn't a more difficult or more richly rewarding transit than this. Remember, Pluto is the planet of life and death. No, this doesn't mean death is rapping at your door—but you'd better believe that some part of you, whether big or small, will go through a massive rebirth during a Pluto transit. It's a transformational passage similar to Uranus, with way less surprise.

Pluto is slow because he is so far away. It can take years to go through one of his transitions, and sometimes it takes longer to figure out what even happened. Deep-rooted change takes time and Pluto wants us to get in touch with the depths of our soul. It's intense and sometimes extreme, but also healing and restorative. Even a hard Pluto transit can be better than therapy.

Many of us are stronger than we know, and you'd better believe Pluto will unearth our powerful capabilities, leaving us with a lasting knowledge of self-sufficiency and fortitude. We will never be the same person we were after Pluto comes to town.

3

THE PLANETS IN THE HOUSES

Ready to add yet another piece to the astrological puzzle? We've learned what it means when your Sun is in Leo, but what about if your Leo Sun is in the Eighth House? Go back to your trusty natal chart and locate where your own Sun lies (hint: it's the circle with a dot in the middle). It will be located in one of the twelve astrological houses. In fact, go ahead and find *all* of your planets—head back to pages 6 to 9 for planetary symbols if you need a refresher! The way the planets are strewn about the Chart Wheel are unique to *you*. Locating each one and in its own house will give us extra insight into how your planets operate.

Remember, the signs reveal *how* a particular planet behaves, but it's the astrological houses that disclose *where* that planet is operating. Think of the houses in astrology like individual environments. They set the stage for each planet, so they can play out their part in your own life. For example, we know that Venus represents love, but depending on the sign she appears in, that love shows up differently. Locating Venus in your natal chart gives us that extra added insight into her motives and direction.

Once we understand how the planet functions, where that planet resides, and who that planet is talking to (aspects), then we are on the way to becoming our very own astrologer!

You'll need your birth time for this one, and a copy of your chart. Once you have that, we're off and running.

FIRST HOUSE

The First House reveals our traits and tendencies right off the bat. This house is on full display for everyone to see, so any planets that reside here are the first responders to any situation you encounter. The closer to the Ascendant, the stronger the response. But really, any planet in this house counts.

SUN

The Sun shows where our personality shines, and if yours is in the First House you've got personality galore! You're a natural leader and need to be in the center of things, too. Make sure other people get to share your spotlight, though. We wouldn't want your popularity to wane!

MOON

You wear your emotions on your sleeve, and you might even resemble your mother, either by looks, personality, or both. Your instincts are incredible, but you can be moody. Don't try to hide your feelings, though—people can see right through to them.

MERCURY

You talk. A lot. You're curious and active when you're at your best, restless and easily distractible when you're not. You're probably a busybody and you respond to life with a million questions. You're so adaptable and spirited, you probably have tons of friends—unless, of course, no one can get a word in edgewise. Remember that listening and rest are important, too.

VENUS

You're gracious and charming and very, very attractive. People gravitate toward you, since you simply ooze charisma. You have a need for beautiful things, and can't stand any scenario that doesn't have a dazzling quality to it. Watch out for being too lazy, though. You don't have to lift a finger when you're *this* magnetic.

MARS

You respond to life with action and loads of it! You're fast and aggressive and have a strong desire to be in charge. Watch your temper unless Mars is in a strong aspect with Saturn or Neptune. Your enthusiasm for life is contagious, and so is your energy.

JUPITER

You've got luck pouring out of your ears! And thank goodness, since you're known to spout some wild tales to anyone who will listen. Yes, you're outrageous, but always a delight. It's hard not to love your effervescent personality and optimism. The bad news is that you tend to overindulge and no one can tell when you're down. You always seem so content.

SATURN

You age in reverse, especially if Saturn is near the Ascendant. You are quite ambitious and responsible but also cautious and full of fear, too. You project wisdom and skill, regardless of your level of expertise. Purpose is your golden ticket. Once you find that, Saturn here will serve you well.

URANUS

You're an odd duck. You stand out and you don't like being told what to do, but your spontaneity makes you a real breath of fresh air. You respond to things in a drastically different way than most people would expect. You shock and awe. Basically, you're a blast! Nontraditional to the core, you're always searching to change things up.

NEPTUNE

You make a fabulous actor, as you can be any character you want to be. This can be frustrating, though, because people can easily project onto you. They just don't seem to really know the *real you*. It's the curse of your many different facades. You're ethereal and dreamy and sensitive, too.

PLUTO

You're secretive and intense. People seem to either love you or hate you as you—no in between—and you respond with a similarly black-and-white perspective to all situations. You have a powerful presence, my dear, so use it wisely.

SECOND HOUSE

The Second House signifies our relationship with our money and our values. Any planets here indicate our self-worth and how we relate to material things. This is the house where we find our personal code of ethics and what makes us feel comfy and secure.

SUN

It's plain and simple when you have the Sun here. You crave security like nobody else. It's your personal mission to focus on what you have, but don't forget to align yourself with strong core values, too.

MOON

The Moon reveals what we need, so having this planet in your Second House means you need money, honey. But just be careful that your finances don't wax and wane like the Moon. And don't feed your emotions with shopping sprees, either.

MERCURY

You like to talk about money, and are probably pretty good with money management as well. You're very practical-minded and may even make your own wealth via communication. But watch out for boring everyone with "just the facts."

VENUS

You have extravagant taste, my dear. With any luck, you'll be an interior designer or personal shopper so you can buy luxuries for others without draining your own savings. But if you're not, make sure your bank account can accommodate your strong sense of style and flair. Possessions may make you feel safe and cared for, but watch out for equating stuff with love.

MARS

You're ambitious when it comes to making money. It's easy for you to get it when you need it, but it seems to leave just as quickly, too. Try to let it linger awhile longer. Material possessions excite you, as well as food and drink. Yep. You're a tad hedonistic, but your values are crazy strong.

JUPITER

The luckiest planet of all in your money house? The green stuff just seems to fall out of the sky for you. As a result, you likely spend quite a bit as well. You're generous and a great gift giver. Just make sure you don't give it all away!

SATURN

Wherever Saturn is in our chart is where we are fearful and insecure, so if you're seeing this planet in your money house, it's an invitation to get to work and focus so our fears of scarcity don't come true. The smart ones realize that self-worth is more important than a bank account, but everyone with this placement sways a bit to the conservative side regarding money. Not a bad thing!

URANUS

You spend and make money in bizarre and erratic ways. You also need a lot of freedom in your finances—that is, not being overly dependent on someone else. This also can denote someone who is happiest freelancing or working in the gig economy. Watch out for being too impulsive when it comes to spending.

NEPTUNE

It's hard to pay attention to your finances when you have the planet of confusion here. But you're incredibly generous and will give anyone absolutely anything they need. You're oh so kind. But seriously, hire someone to manage your bank account.

PLUTO

You obsess over money and many people who have Pluto here actually become quite wealthy. Maybe it's because you're infatuated with the green stuff. In any case, your instincts around your finances are spot-on and you're very protective of your assets.

THIRD HOUSE

The Third House denotes our communication style and the way our brains work. Planets here indicate the way we speak and how we get our point across. This house also rules short-distance travel and neighbors and has a *lot* to say about our siblings!

SUN

The Sun is where we shine, and you absolutely stun in the realm of communication. It's your life's purpose to talk, write, sing, yell, whatever you need to do to get your point across. But remember to listen, too.

MOON

You *need* to communicate your emotions when you have the Moon here. You love to talk and learn and you're quite up to date on the latest news. One of your siblings is probably highly emotional, too.

MERCURY

This is Mercury's favorite place to be! The planet of mental expression in the house of mental expression. You excel at almost all areas of communication and conversation. You're bright and quick and one of your siblings is also a big talker, too.

VENUS

You converse with charm and grace and probably have a beautiful speaking and/or singing voice, too. You have a way with words that spellbind and entrance, like a Siren at sea. You can lure just about anyone in with that gorgeous voice of yours. A podcast may be right up your alley!

MARS

You're a bit forceful with the way you talk and you definitely put a lot of energy into your thought pattern and communication. You may have an aggressive or crazy athletic sibling and your brain never shuts off, either. Meditation is crucial with this placement.

JUPITER

Your words are buoyant, upbeat, and always chipper! Of course, there's a catch. Sometimes you sound so confident that no one knows when you are down and need help. Don't forget to let people know, okay? But otherwise this is a great place for a confident way with words.

SATURN

You were probably quite shy growing up. Insecurity around communication plagues you, but with extra work and effort you can absolutely be amazing in getting your voice heard. Though you are always quite serious in thought, you are very intelligent and honest, too.

URANUS

You say the most shocking things. You stun and awe and have a tendency to put your foot in your mouth, too. Your mind is eccentric and almost ingenious. You're very smart if not a tad bit bizarre and you are highly original in your thinking.

NEPTUNE

Your voice is dreamy and soft and you might be quite artistic or musically gifted. Creative writing is something you excel in as well. Though logic and reasoning sometimes feels out of reach, you thrive in the world of abstract contemplation.

PLUTO

When you speak, your words have power. So much so that you make a fabulous counselor or therapist. If you have any brothers or sisters, you either love them or hate them. Never in between.

FOURTH HOUSE

The Fourth House designates our roots and our foundations. It shows what life was like when we were young and it reveals how we will be in our old age, too. Planets here give clues to what a parent was like, and what traits were passed on from your family. The Fourth House shows the way we nurture and the home life we have created for ourselves.

SUN

Oh, you're a homebody, all right! You thrive in domestic life and are probably a bit of a night owl. And even though you're a nourishing soul, it's the emotional security that you provide that really makes you so popular. That and your stunning home!

MOON

You're sweet and sensitive and oh so moody, but your home is a fabulous place of refuge. Without it, you get sick or all out of sorts. Traditions and family are imperative, but watch out for lingering too much in the past.

MERCURY

You prefer a busy, loquacious household. Of course, you may become more talkative and mercurial the older you get, too. You think in terms of safety and security but you can be highly creative in your thinking.

VENUS

Your home is gorgeous, balanced, and beautiful. Even if you live in a dump, you have a real eye for making it appear exquisite. You're a sweetie, too. Oh-so-sentimental and very affectionate, you'll become quite the charmer in your old age.

MARS

You might have grown up in an aggressive or action-oriented household, so watch for repeating that pattern yourself. You put loads of energy into your home and are usually very protective of those you love. As time progresses, you will probably be very active as you age, so lucky you! Keeping busy and active is a nurturing and restorative activity for you.

JUPITER

Even if your parents were total losers (and they probably weren't), you inherited the best possible genes they could give you. You're probably the favorite child, and you're a fabulous nurturer to boot! You're a generous soul who will do well with plant and fur babies alike.

SATURN

Perhaps a certain parent was undeniably strict, or not around much at all. In any case, you probably didn't have the best experience and you lacked that much-needed warmth and affection. You crave security. It's beyond crucial so you work hard to make sure you never have to worry about that.

URANUS

One of your parents was probably crazy or eccentric, or both! You like to move around and you prefer a home with a definite unique twist. Likewise, you tend to nurture in distinctive and unusual ways. The good news is, though, as you age you won't give a rat's ass what anyone else thinks of you!

NEPTUNE

You definitely felt like you didn't belong with the rest of your tribe. There's something different about you than the rest of your family. You're sensitive and dreamy and do very well living next to a large body of water. Psychic abilities are strong with Neptune here, too. They get even stronger with age.

PLUTO

One of your parents waved the scepter of power and it may have been ever so intense. Never fear, though, as the power definitely shifts to you the older you get. Transformations can take place in your home and you do well to use your abode as a place of beautiful regeneration.

FIFTH HOUSE

The Fifth House is the home of love, self-expression, and creativity. It shows how we like to have fun and shows what kind of parent we can be and also what sorts of children we can have. It rules gambling and sports and games of all kinds. Any planet in this house will reveal not only our dating style but how we behave romantically, too.

SUN

You're highly creative and exceptionally theatrical, too. You have flair that would put RuPaul to shame. Basically, you have epic stage presence, darling! You are here to create, and create you must. You thrive in the spotlight and are a hopeless romantic, but *watch out for too much drama.* Remember, the show must always go on.

MOON

I don't care if you work for the IRS, you have the need to nourish yourself with creativity. Here is a dramatically placed Moon with a desire for fun and romance and all the frivolous things life has to offer. Oh sure, you love to dream and your emotions can sometimes be a bit exaggerated, but truly, you are a kid at heart.

MERCURY

You think in strictly artistic terms and may actually dabble a bit in the creative writing realm as well. You need lots of poetry and romantic conversation from your lovers and if you do have any kids, the firstborn will be quite a talker.

VENUS

There's no way anyone is going to get a date with you if they are not utterly, exquisitely, hopelessly, and delightfully charming. It's only because you yourself are such an absolute delight! You're terribly artistic and creative and a perhaps a tad self-indulgent.

MARS

You are quite the showoff and a very fun date. Proud, confident, and clever with a crafty soul, you have loads of energy—just make sure you find an outlet! Either a sport or something creative. Probably sports, though. And it's highly likely that your children are equally as active as you.

JUPITER

In a nutshell, you're a blast. People should hire you to attend their parties, you're so much fun. You're great with children and you have a big, generous heart, too. You do well in sports and the arts but you can overindulge all too easily as well. Unless other things are conflicting in your chart, Jupiter here suggests that you can also be quite a risk-taker.

SATURN

Self-expression is not something that comes easy for you. In fact—it's a bit terrifying! But the arts seem to run in your veins and as difficult as it may be to let your creativity shine, it's also crucial for your development.

URANUS

Your love affairs are rather shocking and your lifestyle is truly zany. You're a natural creative with a rebellious flair for the outrageous. You need freedom in your dating life and anyone who tries to pin you down will be deeply disappointed.

NEPTUNE

You have an uncanny ability to tap into the creative zeitgeist. Like a personal crystal ball that presides over trends and aesthetics, you simply know what is in before anyone else has a clue. You also love to be in love, even when it's not with the right person. Watch out for deluding yourself every now and then.

PLUTO

There's no such thing as a light and frivolous relationship when you have Pluto in the Fifth. You prefer intense and crazy deep connections. But watch out for obsessiveness tendencies here! Pluto is always all or nothing.

SIXTH HOUSE

The Sixth House is the realm of health, routine, scheduling, and work. It doesn't sound very fun, but it's incredibly productive and any planets in this area assist us in the way we take care of ourselves and our habits. Which is why you'll find your dietary habits and what sort of food you prefer here as well. And since the Sixth House is also linked with small pets, it will reveal what sorts of pets we like.

SUN

With the Sun in the Sixth you are here to show the world what it's like to be of service. Basically, you are here to help! Please don't underestimate how powerful this Sun position can be. Your work habits are like no other and you can be perfect in all that you do.

MOON

Oh, you're a workaholic all right, with a strong emotional need to be a productive, healthy example for all to see. You simply need structure and organization and, let's be honest, you do it pretty well. Watch out for mothering or micromanaging people at work.

MERCURY

You worry, and often. This is the place where Mercury tends to analyze, again, and again, and again. It's a testament to your intelligence and mental stamina, but don't neglect your intuition, either. You tend to favor logic over your incredible instincts, but you need both for real wisdom.

VENUS

You show your love by being of service. You're practical and efficient and your work environment had better be pretty damn beautiful, too. You possess such fine attention to detail and you tend to be very health conscious.

MARS

If you have Mars here, make sure to take a vacation every now and then. Quick and adept, you get things done at an impressive speed. Of course, it's maddening how slow everyone else is, but really, not everyone is as productive as you.

JUPITER

You love to work and it's usually easy to find it, too. It's also easy for you to get promoted, as you are ever so popular in the office. Just watch out for overindulgence in food and drink. To you, the richer the better!

SATURN

Remember, there's working harder and then there's working *smarter*. You were never taught how to do the latter. You believe you have to work for work's sake and flattening your nose on the grindstone is the only way to get ahead. Once you realize you don't have to kill yourself in order to succeed, you'll be much happier.

URANUS

Not just any old job will do when you have Uranus in the Sixth! You get bored and need variety. Drudgery bores you and your organizational skills are, um, different. You seem to have your own managerial system and you prefer doing it your own way. That's okay; it just seems to perplex everyone else.

NEPTUNE

Okay, okay, so you daydream on the job quite a bit. You misplace things easily and find it difficult to follow any strict routine. Best to try to do something creative and increase any volunteer work, too. It's Neptune's favorite thing to do.

PLUTO

The planet of extremes in the Health House means green juice obsessions one minute and a Cheeto bender the next. Gray areas don't exist to Pluto, so try to find balance by establishing clear routines for all aspects of your life.

SEVENTH HOUSE

The Seventh House is usually called the Marriage House, but really it signifies *all* relationships. Any planet that occupies this space reveals what we seek out in another person. In short, the Seventh House will expose what and who we attract.

SUN

You seem to glow in all areas of partnership. In fact, it's your soul mission to be a shining example of what it means to be in a healthy relationship. You probably attract confident and occasionally arrogant people. And watch for giving too much of yourself in any alliance. Just make sure to take care of your own needs, too.

MOON

You crave connection—it's your emotional support! True, you can be a bit needy, but you really treat friends and lovers like family. You give yourself wholeheartedly and feel the need to make others happy, too. You're a natural nurturer.

MERCURY

Communication is crucial to your happiness in any relationship. Silence is excruciating to you and it's necessary to bounce your ideas and thoughts off almost everyone you encounter. True, it's difficult for you to make decisions at times, but you are the best at seeing every side of a situation.

VENUS

Oh, you're a real charmer, all right, and your relationships are equally delightful. You love to be in love, and you can't stand being alone. Good thing for you it almost never happens! You're far too captivating to be isolated for long.

MARS

You like a challenge in the relationship department and don't mind at all a little bit of competitiveness, either. You're passionate when it comes to love and you seem to attract active types and sometimes slightly argumentative people. But remember, you're just as strong as they are. If not more so at times.

JUPITER

You're elegant and alluring and seem to attract the best of the bunch! Because you are naturally charming, people flock to you like the Dalai Lama. You find yourself lucky in love, but watch for overindulging in the romance department, too. Jupiter can always be excessive.

SATURN

Relationships terrify you. Maybe because you didn't really have a healthy example of what a good partnership looks like. Whatever the case, you seem to find marriage a very serious affair. So serious, in fact, you demand perfection before you even decide to put a ring on your finger. There's nothing wrong with taking your damn sweet time, but remember that nobody is ever perfect.

URANUS

You tend to attract the strange and unusual. It's just as well, as it's freedom that you really want. No boring, conservative love life here! Friendship before marriage is equally important, and you also need loads of spontaneity.

NEPTUNE

Yes, you lure in the romantics and creatives, the drunks and the cheaters, too. Every now and then, be sure to remove those rose-colored glasses and ask a friend for a reality check. In truth, you can inspire people. Just make sure they are deserving of *you*!

PLUTO

You need depth in your relationships. Nothing superficial for you! The tall, dark, and brooding make you drool but do watch out for power struggles, too. All of your connections are intense and ever so passionate.

EIGHTH HOUSE

Sex and other people's money define the Eighth House. This is classically called the "Merging House" and death, taxes, the occult, and psychic abilities reside here. Any planets here divulge how we are in the bedroom as well as what our partner's financial situation looks like. This is also the rehabilitation realm. It's a powerful and profound house!

SUN

You know a thing or two about power, don't you? Basically, you know how to harness your strong transformational abilities and in turn you can assist others in this magical process, too. You're deeply healing when you want to be, and you're blessed with amazing psychic abilities as well.

MOON

You have a strong, deep need to dig below the surface of life. You crave profound experiences and even stronger emotional connections. You are simply fascinated by what lies below the surface of people and situations.

MERCURY

You're phenomenal at uncovering hidden secrets and motives. And you're smart. With a detective-like brain and strong intuition, you can suss out any insecurities, truths, or sneaky agendas. But do watch your forked tongue! It can be quite deadly at times.

VENUS

No casual love affairs here! It's intensity that you seek. Well, that and the nicest things life has to offer. You need your partner to provide security and a passionate, vibrant sex life. It's crucial for your happiness.

MARS

Intensity is your middle name! You're powerful, profound, and strong, too. Trust me, no one wants to mess with you. You're a powerhouse in the bedroom and your intuition tends to be so dead on that it's almost spooky.

JUPITER

You're a healer and a mystic with a strong sense of what is right and wrong, though this never stops you from having fun! Your motto is "go big or go home" and you do. You live life fully and beautifully and you don't take things for granted.

SATURN

You know you want it more than anything. That's right, I'm talking about deep, intense, emotional intimacy. It scares you to death. Well, that and your own power. You weren't really ever allowed to wield it. Learning to surrender yourself and being in total control: that's your lesson in this life.

URANUS

You adore change and transformation. Life would be insanely dull without it! You're also quite kinky in the bedroom and need heaps of personal space. You also possess a different and unique perspective when it comes to death and the taboo. Nothing is bizarre to you.

NEPTUNE

The most psychic of the bunch! Your eerie insight is incredible. You connect with worlds that most people aren't even aware of, and your sex life is equally rich in fantasy, too. Just remember that boundaries are something you aren't fond of. But make sure you respect others.

PLUTO

You have a very intimate relationship with your impressive metamorphic powers. Within yourself *and* with others. This, of course, makes you a fabulous therapist and healer and you definitely don't shy away from the occult. Your love of all things dark and mysterious is definitely contagious.

NINTH HOUSE

The Ninth House signifies travel and higher education. It's also the Teaching House and it hosts the realm of major mind expansion. Planets in this house reveal where we prefer to plan our excursions, what our life philosophies are, our beliefs and values, and our ability to perceive right and wrong.

SUN

With the Sun in the Ninth House it's your general purpose in this lifetime to provide intellectual growth and share insights with the rest of us. You're a fantastic teacher and your love of travel and adventure is necessary for your development and that of the people around you.

MOON

You nourish yourself with travel and learning and your principles are often based on emotions rather than logic. Your advice is golden and your perspective on life is something we all wish we could have. This is a fabulous place for the Moon.

MERCURY

You have a cheerful outlook on life and make a fabulous teacher or salesman. You have the very unique ability of seeing everything with a bird's-eye view, so you are always happy to air your opinion.

VENUS

You have a real love of adventure, but if your destination isn't beautiful, you'll get depressed. You also probably prefer your love interests to be foreign or to come from an almost completely different background than your own. And you're very charming and persuasive when you talk about ethics and your beliefs.

MARS

You can get quite heated during any debate, which you always seem to find yourself in the middle of. You put your energy in spreading the word, and when you think you're right, I truly feel sorry for the other guy. You'd make a fabulous attorney.

JUPITER

This is Jupiter's favorite place to be! You're lucky when you travel and with respect to learning. You may have even been the teacher's pet. You have a big, generous heart and are a true bohemian at heart.

SATURN

Your horizons were somewhat restricted in youth, either by travel, or education, or simply what you were taught to believe. But you get to make your own rules now and, by your own example, you're here to illuminate a path to self-discovery for the rest of us.

URANUS

Whatever you were taught to believe in your youth is the exact opposite of what you believe in now. You really do think for yourself! You may have unusual or unpopular opinions, but you aren't afraid to express them. You also love to travel to any strange and exotic destination, and you possess strong (if untapped) telepathic abilities.

NEPTUNE

You fantasize about mystical, strange, and foreign lands. You might even be into sci-fi and high fantasy—anything that will transport you to another world. You're a real dreamer and when you travel, please make sure it's near water. It's incredibly healing for you.

PLUTO

You possess powerful urges to search for deep meaning in everything that you do. You probably have been through a few philosophical transformations in your lifetime and you strongly believe in magic and the occult. Knowledge is power, and you understand this more than anyone.

TENTH HOUSE

This is the realm of career and public recognition. And, similarly to the Fourth House, the Tenth House denotes one of our parents—in this case usually the mother. This house exposes how we behave relative to authority and people of power. And any planets in this house will give us clues to how we want to make a living, and what we'd like to be known for.

SUN

With the Sun in the career realm, you are here to be a shining example of ambition and worldly success. Easy, right? Well, you *were* born when the Sun was shining midday. This is why everyone is always in your business. You can't hide a thing, so go ahead and step into that spotlight you so very much deserve.

MOON

You need success, it's true. Fame is what you want but not only for fame's sake alone. You want recognition for all of your hard work. You're sensitive to what the public needs and you are more than eager to provide. And don't neglect your amazing intuition around your career—follow your gut!

MERCURY

You're able to apply your amazing communication skills in the realm of work and recognition. Perhaps you're even well known for your voice or ability to talk to just about anyone. You're an excellent writer, too. Just make sure to always keep up good conversation between you and your boss.

VENUS

You probably do something artistic or beauty related, but even if you don't, you make everything you do look fabulous. Venus always reveals what we love and you, my dear, love your career.

MARS

You put a great amount of energy into your career, and have a better than average chance of becoming a serious workaholic. It's what you do to get ahead, and boy do you get ahead. Just watch out for arguing with your boss too much. Even better? Become the boss.

JUPITER

Is it any wonder that everyone keeps trying to promote you? You're the favorite in your field! Always chipper and spreading enthusiasm to everyone you encounter. Basically you just have to show up to work and everybody loves you. Just don't get complacent—luck is good, but skill and dedication are important, too.

SATURN

Someone instilled in you a great deal of fear around being a success. Oh, you're a terribly hard worker, all right, with too little praise. Keep working, though. The praise that comes from within is far better than anything that Mom or Dad could give. And success will definitely come when Saturn is here.

URANUS

I have to be honest, you make a terrible employee. You loathe authority and can't stand being told what to do. So what's your solution? Be your own boss and do your own thing. It's really what you crave anyway.

NEPTUNE

You prefer a career in the arts, as you're incredibly creative. You also love to help people. Healers and artists make up the majority here. And don't worry about not having a concrete idea of what you want to be when you grow up. The idea of a "profession" is a fluid abstraction.

PLUTO

Power, baby. That's what you want. That and control. You have an insatiable appetite for success and you'll usually get it, too. Just make sure to use it wisely: your powers lie in your ability to transform, not to dominate.

ELEVENTH HOUSE

The Eleventh House is the place of friends, groups of people, and community. It's also the realm of future planning. Moreover, any planets that take up residence here indicate your social circles and also reveal your hopes and your wishes.

SUN

You're a team player with strong leadership skills. The way you shine is by leading groups of people, fighting for a cause, and embracing societal dreams. And there's no way you can do that with an iron fist. You rule by being a cog in the wheel, all the while expressing your uniqueness and inspiring others.

MOON

You mother your friends and in turn you feel incredibly nurtured when you are among your favorite groups of people and associates. Your dreams for a stable society are intense, and you can help foster that by utilizing your marvelous instincts around how to make the future a better place.

MERCURY

You're attracted to people and groups that hold your same mind-set and point of view. True, your friends are talkative, but you probably have a variety of social circles that you flit in and out of. You think in terms of the future, not the past, and you're an intellectual at heart.

VENUS

You make a fabulous hostess, as you can charm and beguile like no one else! You have a beautiful friend group, in both looks and manner. You may even find that you prefer your friends to your lovers. Bonus if you can snatch one that can be both.

MARS

Here's an activist who will fight to the death to defend what is right and what is wrong. Social justice is big to you and you jump at the chance to march in any rally or protest. Your friends are just as feisty and you all love a good, heated debate.

JUPITER

Your friends are undeniably beneficial for you. How can they not be, as you seem to pick the best of the bunch?! You thrive in all forms of social situations and people seem to be your magical good luck charm. Good thing you're lucky for them, too.

SATURN

You always have a sneaking suspicion that you've never quite fit in. It's what makes you special, though you'd really like to be like everyone else. You have a unique lesson in this life to be a great social contribution, whether or not you look or act like the rest of us.

URANUS

You definitely have unique and odd ways of improving this world, though your real talent is inspiring others with your battle cry. Yes, your friends are eclectic, from every walk of life and from every age, too, but your open-mindedness is a gift.

NEPTUNE

It's true you'd love to save the world. It's a painful place and you have high aspirations and dreams of a better tomorrow. Good thing most of your friends are artists and dreamers themselves. But do watch out for the sneaky ones, too. Oh, you attract them all.

PLUTO

You have powerful and intense friendships, and your hopes and dreams are equally as passionate. You don't like too many people in your friend circle, though. Best to have a select few that you can engage in deep conversations. Superficial small talk bores you.

TWELFTH HOUSE

The Twelfth House is the most spiritual house of all. It's the home of secrets and your unconscious. It's the realm beyond the veil and any planets that reside here show us where we have faith and what we'd prefer to keep hidden.

SUN

This is a creative and otherworldly placement to have the Sun. You're psychic, artistic, and dreamy. But you can't come fully into yourself until you learn what you are *really* here to do. Devote your life to the emotional service of other people. It's truly where you shine. Once you master that, you'll get the attention you truly deserve.

MOON

In the House of Hidden Things, it's hard for people to know exactly what you are feeling with the Moon here. It may even be confusing to you at times. But you are definitely imaginative and very, very intuitive. Learning to evoke boundaries is so good for you.

MERCURY

You're naturally introspective and very private when it comes to voicing your concerns. It's not easy for you to stand up in a crowd, but when everyone finally hears your uncanny insight, you'd better believe they will listen.

VENUS

You're sensitive and soft, and your love runs insanely deep with limitless boundaries. You prefer people not to poke around in your love life, or your music, or what you like on Instagram. In short, you're private!

MARS

It's a strange place for Mars in a chart, but it does happen! For one, you don't always show your anger to yourself, or to anyone else. Then it comes out of nowhere and you scare the crap out of everyone around you. Finding healthy outlets for your emotions is essential for your well-being.

JUPITER

You're a sweet soul who would give the shirt off your back to anyone in need. The Universe really seems to have your back—it's almost like you're protected by unseen things. It's a good thing, though, since you spend so much time trying to help others: someone's gotta be looking out for you!

SATURN

Faith is not your strong suit. Well, it never was. No one ever let you believe it was ever there in the first place! Forgiveness is crucial to your development, as it's your lesson to be of spiritual assistance to others—after you, of course, learn how to heal yourself.

URANUS

You're a secret rebel! No one knows what's really going on behind the scenes with you. They'd be shocked, truthfully. But you're far too rebellious to actually care. You're a weirdo beneath the surface and when you finally let your freak flag fly, you feel liberated.

NEPTUNE

This is Neptune's favorite place to be! Your intuition and psychic abilities are strong and you're incredibly creative as well. You're able to spin your dreams into art, but you never get credit for all the amazing work that you do. You're far too humble to notice.

PLUTO

You're intense and powerful, but it's not easy for you to flaunt it. In fact, you sometimes actually repress it. Then, like Mars, that power comes out in strange and inopportune ways. Learning to embrace your shadowy side is crucial to you.

4

ECLIPSES AKA COSMIC CHANGE AGENTS

Ah, the dreaded Eclipse. It always seems to scare the bejesus out of people, and while there's really nothing to be afraid of, they are pretty eerie. I can't imagine what people were thinking when they first saw the Sun slowly shift to black in the sky, or watched the Moon become red. That's some pretty apocalyptic stuff right there! Not to mention the crazed emotional state every Eclipse seems to leave in its wake. Feelings are volatile, people become hysterical, and the world feels a bit unstable.

In astrology, Eclipses bring change. And while change is necessary, we humans prefer our routine and predictability, thank you very much. Eclipses stir us up. Eclipses occur four, sometimes six times a year and happen about every six months, and every time they bring up feelings and emotions that we could've never imagined were there. They shift our perspective and unstick events and situations that have become stale or stagnant. Eclipses are a little bit like a colonic. They'll get things moving. Fast.

A Lunar Eclipse occurs when the Earth moves between the Sun and the Moon, blocking the light to the Moon. This is similar to a Full Moon, but much more intense. A Solar Eclipse happens when the Moon stands between the Sun and the Earth. It blocks out the Sun, and is basically a New Moon magnified. A Solar Eclipse is associated with fresh starts and new beginnings, while a Lunar Eclipse is more in line with endings and full completions.

You'll need an ephemeris (a book or table that configures future planetary positions) or an astrologer to tell you when the next Eclipse occurs. By the way, an ephemeris is a book that shows us where the planets and asteroids are every day of the year. Get one, and look back on all the Eclipses that have occurred during your lifetime. You'd better believe that one of these suckers lit up your chart during some of your biggest life events. It's pretty amazing stuff, actually.

Eclipses can bring news of a new job, marriage, or baby. You'll find people making major life decisions during this time. Some people move across the country, and some get a divorce. Some do both! Of course, it absolutely depends on where the Eclipse is materializing in your own chart. Extra points if it happens on one of your planets, too!

An Eclipse is also like a swift axe. It will sever relationships and situations that are no longer viable to you or in line with your life's direction. It's a revealing time, and you'll get insight galore into a person's motive or an event. It'll give us major clarity, so don't ignore this.

Paradoxically, the best way to handle these astrological events is to not try too hard. Look, we're already pretty crazed about two weeks prior and three days after an Eclipse. No need to get our adrenaline on overdrive here. An Eclipse is a massive turning point, whether we like it or not. There's not much you, or I, or anyone else can do about it. Count three days after an Eclipse before you act on any of those life-changing realizations. We're all a bit saner, a few days after an Eclipse occurs. Meanwhile, sit tight and try to relax. Here are my tips for dealing with these crazy times!

SOLAR ECLIPSE SURVIVAL GUIDE

Solar Eclipses are a time to start something new. These periods might amp up drama in your life or open doors to transformational opportunities that will take courage to embark upon on. Stay open to possibility and welcome the unexpected! Life-altering changes might feel exciting or downright terrifying, but I've got your back. To cope with these periods of unanticipated shift, extra self-care is a must. Eclipses are a good moment to . . .

LAUNCH A NEW PROJECT OR HOBBY. Start that book you've dreamed about! Download the language app. Take up bread making.

CLEANSE YOUR HOME AND YOURSELF—spiritually and mentally. Focus on an area in the home that you might normally neglect—behind the fridge or the baseboards, or donate shoes and jackets gathering dust.

PLANT NEW SEEDS, literally and figuratively. Get out in your garden and get your hands dirty. No yard, no problem! This is a great time to invest in some house plants.

UP YOUR SELF-CARE! Plenty of sleep and healthy food will prepare you for your new beginning. Treat yourself to a silky pillowcase and try new salad recipes.

JOURNAL TO GET CLEAR ON YOUR INTENTIONS. Make a list of five goals each morning.

SAY NO TO ALCOHOL AND OTHER ESCAPIST TENDENCIES. Staying grounded is crucial during these times, and distraction is not the cure. Sip a new flavor of tea instead.

INTENSIFY YOUR WORKOUTS! Seriously. Breaking a sweat is hugely beneficial for all the anxiety swirling around in the air.

LUNAR ECLIPSE SURVIVAL GUIDE

Lunar eclipses are moments of release. This is a wonderful time for self-reflection as you prepare to let go, while also working to remain open for something new to take its place. These are ideal moments to . . .

KEEP AN EVEN TEMPER. We're more emotional during Lunar Eclipses than Solar ones. Compromise as much as you can and resist emotional outbursts.

JOT DOWN A LIST THINGS YOU'D LIKE TO RELEASE. Consider burying or burning it safely to set your intentions.

MEDITATE and spend time alone.

GET EXTRA SLEEP. Insomnia runs high during Lunar Eclipses. Go with it and don't stress, but lavender salt baths are amazing for this.

KEEP A DREAM JOURNAL. The Moon brings clarity around your subconscious, and the dreams you have during a Lunar Eclipse can be magically revealing.

DECLUTTER your home, closets, or workspace.

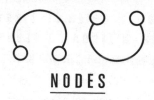

NODES

Nodes aren't planets or asteroids—they're mathematical points where the Moon and Earth intersect. Interestingly, Eclipses always correspond to the Nodes in the sky: when a Full or New Moon is 19 degrees near the Nodes, we have an Eclipse!

Everyone has a North Node and a South Node and, as you might expect, they always sit opposite one another. For example, if you have the North Node in 11 degrees Aquarius, you'll know that the South Node will land in the opposite sign of Leo at 11 degrees.

The North Node represents what you've never done in this lifetime, or any of your past lives, either. It's where you'll grow and evolve and discover your true calling. It's your karmic destiny! It's where you're lucky. Also, if someone has their Sun on your North Node, they're here to assist you in activating your life's mission. The North Node is your bliss. Learn it, and try to do it well.

The South Node represents what we are bringing from our past. We've done it thousands of times, and if we keep it up, it's only going to keep us stuck and complacent. Sure, it's comfy and easy, but it isn't going to do us any favors if we keep repeating the same behaviors and goals. We can't expand if we remain in our comfort zone. Don't mess with too many people who sit on your South Node! They'll only hold you back.

The goal is to manage both North and South Nodes equally. Use the South Node as a springboard to learn your North. After all, we can't completely neglect all of our past lessons. We have to bring some of it forward to assist in our new direction, too.

Any Eclipse that falls on your North Node is instrumental in aligning you on your path, while an Eclipse on the South Node indicates a release, or letting go of whatever is holding you back. Take a look at the Nodes below (and compare them to your personal natal chart) to get a sense of where you are headed in this lifetime and what it is you should surrender.

NORTH NODE IN ARIES OR FIRST HOUSE / SOUTH NODE IN LIBRA OR SEVENTH HOUSE

You have relied on others' opinions, or you have learned to see yourself through the eyes of others, for far too long. This is your lifetime to branch out on your own and master self-reliance. Learning independence and assertiveness is the name of the game. This does *not* mean you can't have a beautiful, committed partnership! But you must grasp the knowledge to stand strong on your own; otherwise relationships can be unfulfilling.

NORTH NODE IN TAURUS OR SECOND HOUSE / SOUTH NODE IN SCORPIO OR EIGHTH HOUSE

This is your lifetime to cultivate patience and boundaries. You are learning personal values and the joy of earning money on your own. You have relied too long on the resources of others and it has become deeply unsatisfying. Learning to become more stable brings joy and gratitude. Also unimaginable pride at becoming self-sufficient!

NORTH NODE IN GEMINI OR THIRD HOUSE / SOUTH NODE IN SAGITTARIUS OR NINTH HOUSE

This North Node is here to develop listening skills and communication. You're way too good at spouting off thoughts and beliefs without a care in the world, or taking other people into consideration. You may seem a little self-righteous and preachy, and in this lifetime you're here to see through others' eyes and viewpoints. Listening is your key to success and happiness. Understanding others brings balance.

NORTH NODE IN CANCER OR FOURTH HOUSE / SOUTH NODE IN CAPRICORN OR TENTH HOUSE

Yes, you've been a big shot before. You may have neglected the importance of family and friends in order to achieve success. In this lifetime you are learning to develop a nurturing, sympathetic understanding of others in lieu of fame and prosperity. Home in on your own feelings and sensitivities. Success can be yours but only after you validate others and learn to appreciate your own special tribe.

NORTH NODE IN LEO OR FIFTH HOUSE / SOUTH NODE IN AQUARIUS OR ELEVENTH HOUSE

You're most comfortable in a group and have a hard time learning to shine on your own. But shine you must, and self-expression and creative understanding are key to this Node. You are here to be an original! Stop relying on your friends to dictate your decisions. You don't need to belong to any particular group in order to cultivate your own heart's desires.

NORTH NODE IN VIRGO OR SIXTH HOUSE / SOUTH NODE IN PISCES OR TWELFTH HOUSE

Details are the name of the game here and you'll want to focus on every single one of them with your North Node in Virgo. You've spent lifetimes in a vague, dreamy world filled with escapists and con artists and now it's time to discern who really needs help and who is simply a distraction. You're here to work, stay busy, and remain as healthy as possible. You want peace, and being of service is the way to find it.

NORTH NODE IN LIBRA OR SEVENTH HOUSE / SOUTH NODE IN ARIES OR FIRST HOUSE

You've been on your own for far too long. People, who needs them! Well, you do if your North Node is in Libra or the Seventh House. Learning to cooperate is tough, but it's the way to spiritual and emotional fulfillment for you. Learning to share and balance with others, and leaving behind an arrogant disposition, will unlock a beautiful world with way more possibilities.

NORTH NODE IN SCORPIO OR EIGHTH HOUSE / SOUTH NODE IN TAURUS OR SECOND HOUSE

Transformations and major upheavals are things you'll learn to lean into in this lifetime. Rigidity and stubbornness come all too easy, but you'll soon find that they only hold you back. You get comfy. *Real* comfy and refuse to budge on almost anything at all. It takes mountains to move you and learning to let go can propel you to success. And let go of "stuff." There is nothing material on this Earth that can properly define who you are.

NORTH NODE IN SAGITTARIUS OR NINTH HOUSE / SOUTH NODE IN GEMINI OR THIRD HOUSE

Get out of your neighborhood and travel the world. Your perspective needs broadening and so does your intuition. You spend a great deal of energy making sure you have all the facts with little care for what it all means. Learn to trust yourself, and for God's sake stop gossiping! Freedom and adventure await you. You have a higher calling, so make sure you embark on a spiritual quest, too, every now and then. Your soul needs it!

NORTH NODE IN CAPRICORN OR TENTH HOUSE / SOUTH NODE IN CANCER OR FOURTH HOUSE

You've spent a lot of time behind the scenes, and in the comfort of your own home. It's where you feel safe and secure and why mess with that? But with the North Node in the shining sign of Capricorn, you are here to focus on career, responsibility, and ambition. You can't always retreat when you feel threatened or you'll run the risk of not making it in this world. And that's exactly what you are here to do: succeed.

NORTH NODE IN AQUARIUS OR ELEVENTH HOUSE / SOUTH NODE IN LEO OR FIFTH HOUSE

You're dramatic and prideful and enjoy being onstage. Lifetimes of thriving in the spotlight have rendered you royal, but this time around the focus ain't on you. Applause may be what you crave, but your soul is here to help the group. Focus on friends and assist them with their own creative efforts and you'll be in a win-win scenario.

NORTH NODE IN PISCES OR TWELFTH HOUSE / SOUTH NODE IN VIRGO OR SIXTH HOUSE

Obsessive and critical mind-sets come naturally to you. You've helped so many people in so many lifetimes, you have an instinctive "order" to your methods and a strong desire to be in control. Of course, this now holds you back and keeps you from thriving, so what to do? Let go, relax. Trust in a higher power and learn to develop your intuition. It's stronger than you know.

5

RETROGRADE REAL TALK

Oh Retrogrades. They do get a bad rap—especially poor Mercury. But they really need not be feared! When a planet appears to shift backward, or go into Retrograde, it's simply the Universe's way of telling us to slow down. Proceed with caution. And make sure we're crystal clear about what we are about to embark upon. Working with the Retrogrades instead of against them is key to having a happy Retrograde!

Retrogrades happen when a planet shifts "backward" in its orbit. This happens for every planet with the exception of the Sun and Moon. Of course, the planets don't literally move backward. It's an illusion. Basically, each planet has a different speed and when a particular planet is closer to Earth, that planet appears to slow down its orbit. So slow, in fact, it seems to be spinning backward. This magical planetary proximity is what we call a Retrograde.

Astrology is concerned with how the planets look from Earth so, illusion or not, if the planets look like they're backing up, then for all intents and purposes, they're backing up. And you'd better believe we feel it! We even feel it when a planet is in its Shadow Phase (the wonky time before a planet moves backward and again when a planet shifts forward, or direct).

BORN UNDER A RETROGRADE?

While we tend not to think about Retrogrades when we read our natal charts, it's entirely possible that you have one of these unique planetary movements represented. Take a peek. Do you see any planets with "Rx" next to it? If yes, then you were born with those planets in Retrograde.

When you are born with a Retrograde in your chart, the planet or planets have a strange but beautiful way of expressing themselves. It's often a sign of intelligence, even though that planet is often experienced with hesitation and sometimes fear. You feel it and think it compulsively. It's a bit difficult to express a retro planet, so you go inward. You don't just act and move on. You ruminate on that particular planet, by sign and by house. You know that planet well. It's felt internally *and* intelligently.

People who were born with a Mercury Retrograde are often quite shy when they are young. This retro planet reveals early speech delays and impediments that yield to a deeply penetrating mind later on. You see things in abstractions and not just at face value. You'll notice things most people don't, but you can also project your own thoughts when not necessary as well. These folks are introspective, creative, and quite perceptive. A lot of very successful people were born while Mercury was in a backspin. Steve Jobs, Lady Gaga, Angelina Jolie, and Robert Downey Jr. were all born when Mercury was Retrograde.

If you were born during a Venus Retrograde, you have rather interesting views on how to give and receive love. These folks are serious when it comes to the light and playful side of Venus. Superficial small talk and social grace don't come naturally to them, and they don't trust it when you do it, either. "What's really going on here?" they wonder. "What's their ulterior motive?" People who are born during a Venus Retrograde have a very hard time of trusting. The key is learning to love themselves, something a parent probably never showed them how to do. Love yourself first, and then you'll attract the right kind of people who love you in return. Charlize Theron, Courtney Love, Kate Moss, Ellen DeGeneres, and Jack Nicholson all were born during a Venus Retrograde.

If Mars has a little "Rx" by it, then you were born when the Action Planet was in reverse. This isn't a sign of weakness, but rather an indication that all that action and fighting energy has been turned inward. Anger can feel too difficult to express, so these people bottle it up and let it simmer deep inside. Then months, or even years later it erupts without a warning, and watch out! These folks can explode. You see, they loathe it when other people lose their cool. So unrefined! So uncouth. It's embarrassing to a native Mars Rx, so they learned to repress their own anger, with sometimes disastrous results. It's absolutely crucial for them to learn to deal with their animosity in the present time. Journaling is helpful, as it assists them in connecting deeper with their fears and competitive nature. Morgan Freeman, Dolly Parton, Sarah Jessica Parker, and Carol Burnett were all born with Mars in Retrograde.

The Outer Planets move in a Retrograde motion around five months out of the year, so lots of people are born with Jupiter, Saturn, Uranus, Neptune, or Pluto in Rx.

Jupiter Retrograde people are lucky. They'll snatch up a golden opportunity before anyone else even senses it's there. Funny thing is, just as the rest of us have given up on a difficult scenario, Jupiter Rx folks keep trying and usually persevere. They create their own fortune, and appreciate you by what's on the inside, not the outside.

Saturn Retrograde natives hold on to their fears and insecurities more than the rest of us. They fear rejection and they work hard to make sure that that fear doesn't come to fruition. They do this either by becoming a hermit or by making great achievements. Clearly the road to loneliness is not the high road here. Saturn Retrograde folks have the ability to be quite strong as long as they learn that the occasional bout of misfortune is sometimes simply a part of being alive.

People born under a Uranus Retrograde hide their weirdness well, even though they may be the strangest of the bunch! These guys act like they'll do everything you ask. They'll follow the rules, they'll listen to authority. Don't believe it. They are true rebels underneath it all.

And, sure, Neptune Retrograde people are more gullible than the rest of us, but their rich inner realm is more fanciful than you and I could ever conceive. I often think these folks' dreams must be more real, more tangible, and more extraordinary than we could ever imagine. I may even be a bit jealous.

Born under a Pluto Retrograde? A lot of people have this and yep, they have issues of control, power, and deep-seated change. Some of them have a hard time being in control, though they want it badly. Others flaunt it so much it's flat out difficult to be around them. Basically, a Pluto Retrograde makes it challenging to get in touch with their Plutonian power.

RETROGRADE TRANSITS & HOW TO DEAL

When the transiting planets up in the sky start doing their own backward dance, things get weird. People change their minds and their hearts. Miscommunications and misunderstandings happen . . . a lot. Things break, get lost, reappear, break again. Is it any wonder why people get so afraid of these silly little Retrogrades? But really it's not so bad, as long as you know how to use a Retrograde for your own advantage. The key is to remember the "RE" in everything: REimagine, REdo, REassess, REmake, REassign, REvise, REpeat, and so on. Think of each Retrograde phase as a gift of slowing down and looking within. It's a beautiful period of withdrawal, laced with incredible insight into our own motives and future preparations. We simply can't constantly fling ourselves forward without much thought or consideration, right?

The Retrogrades offer a magnificent period of time for self-reflection. We're able to firm up our foundations and fundamentals for a stronger tomorrow. A time to RE-assess and prepare for the day those planets are moving forward again.

They are also wonderful periods in life to close up old projects and reconnect with old friends and lovers. Lingering arguments and quarrels can be exceptionally healing when the Universe offers us a chance to revamp previously sour scenarios. They can also be an incredibly auspicious time to rethink addictions and routine. I see a lot of folks quit sugar, alcohol, bread, cigarettes, etc. They'll start working out and rethink their personal commitments. Letting go of things is a beautiful way to handle Retrogrades.

THE INNER PLANETS

Because of their proximity to Earth, the Inner Planets station retrograde more frequently and we feel their effect far more strongly. True, they don't take as long as, say, Pluto does in its Retrograde. But these shorter inner planetary Retrogrades sure can pack a punch! Read on.

MERCURY

Mercury is the most common and well known of all the Retrogrades. He moves backward three or four times a year for three weeks at a time and almost always brings a suspension of conversation, information, and delays. Packages get postponed and letters go missing. Travel plans go awry and setbacks are more common than usual. Contracts and agreements seem to be missing key points of information, so a big no-no is to sign anything at all. We overlook things more easily when Mercury is in Retrograde and that may be what the new job entails or how much we'll actually get paid. Don't close on a house or sign a new rental agreement, either. You thought pets were okay, but they're not. Home inspections gloss over major foundation problems or the fact that you'll need to completely rework the plumbing.

Now, sometimes it's impossible to wait it out and if you get offered a new job, or must sign a contract, learn to expect the unexpected when Mercury goes direct. It's crucial to double-, triple-, and quadruple-check *everything*. But let's be clear: if you can, wait. *Wait.*

A lot of times, a better option appears when Mercury is forward again.

Watch your tongue, too, when Mercury slows down! People say things they shouldn't, and sometimes say too much. We often become forgetful during these times and that includes remembering not to spill the beans or the fact that you swore not to tell that very important secret. Misunderstandings are one of the biggest problems during a Mercury Retrograde.

This is the Trickster Planet and techy stuff often goes haywire, too, so back up any documents or devices for added assurance. Don't buy any techy stuff, either, since, weirdly enough, devices purchased during these times seem to be faulty and full of imperfections.

You're clever if you use this time to clean out drawers and closets. Get in touch with old friends, and retackle old projects. Rest, go inward, and exercise. Mercury-ruled Gemini and Virgo often find this Retrograde extra challenging and this includes where you have these signs in your own chart. Gemini on the Money House? Curb your spending stat! Virgo on the Career Cusp? Take a vacation from work. Check where your natal Mercury is, too. All of these, plus where Mercury Retrograde is transiting your own chart, will reveal your own difficulties with this Retrograde. Put these areas on the back burner, so to speak, and you can absolutely thrive during a Mercury Retrograde.

VENUS

A Venus Retrograde occurs about every eighteen months and lasts around forty to forty-three days. Venus is the planet of love and beauty, and when she goes backward, so do our tastes and our desires. We reassess who we love and what we value. We rethink how we dress, how we style our hair, and even reimagine our home decor. Sometimes we see with more clarity, but usually when Venus slows down, we are more delusional, too. This is not the time to make any major permanent beauty decisions. So no to Botox, that expensive dress, or that crazy new hair dye. Wanna repaint your living room? Great, but wait. Are you drawn to that metallic leopard-print couch that never caught your eye before? Yeah, no. *Wait*. Style gets weird during Venus Retrograde. This is an experimental time with aesthetics, but it'll make you cringe when Venus goes direct.

Don't start dating during a Venus Retrograde and *definitely* don't get married. That ex may look oh so appealing, too, but dear God, just *don't*. People break up during Venus Retrogrades and old friends and lovers reappear. Just because our hearts take a trip down memory lane, that doesn't mean our true soul mate was the one that broke our heart. Venus can help us with closure. She's healing, and old flame issues that get resolved only assist us in our future and current relationships. The good relationships stay strong, though Venus will happily test us all.

Venus is all about charm and grace and we see a lot of that go out the window when Venus is backward. People are rude, and tell it like it is. It takes extra effort to form compassionate and tender thoughts and actions. People get gross and they don't care who hears them belch or who sees the stains on their clothes. No wonder so many relationships don't make it past a Venus Retrograde!

Venus also rules money, so no to spending or any big investments. It's more than a good idea to focus on your finances and get them in tip-top shape during a Venus Retrograde. Go through your closets and donate old clothes, clean out the muck in your house and in your love life. Reassess your feelings and your affections. Taurus and Libra are most affected by Venus Rx. Where you have these signs in your chart is where you are affected, too.

MARS

Mars goes backward every twenty-six months and this lasts roughly about ten weeks. The Planet of Anger and Aggression going inward means less outward hostility, right? Wrong. People get mad, frustrated, and irritable. People yell more, honk their horns, and flip everyone off. We question our direction and our goals. Forward momentum stalls out and so does our energy. We're tired and cranky and discouraged to boot.

Don't start fights and don't get surgery when this warrior energy goes backward. Mars is associated with sharp objects, so no to tattoos. Don't start anything new if you can help it, whether that be a job or a relationship or an argument. Initiators lose during a Mars Retrograde, and that includes taking someone to court and filing for divorce.

Mars brings up old rage and resentments. Don't be surprised if feelings of fury flood your veins and irritating memories come back to haunt you. As Mars drifts slowly backward through the skies, old visions and angst come bubbling to the surface. It's sometimes painful, but often incredibly healing. It's best not to shove those memories or issues back down again. Think of a Mars Retrograde as *the* time to tackle painful recollections so we can move forward, lighter and much freer. Mars moving backward can lighten our load.

Exercise is imperative during a Mars Retrograde. It's the healthiest way there is to release your rage. Move your body, exhaust yourself, and don't fuel up with crap, either. Mars wants us to take care of ourselves for healthy productivity. Discovering restorative outlets can make a Mars Retrograde much more bearable. Just don't overtax yourself here.

And remember: Mars is the planet of *action*. Now is not the time to blindly move forward. Our plans are temporarily frozen for good reason. Are we absolutely sure our present path is the correct one? This is a time to reposition ourselves and find the most intelligent ways to use our energies and desires for future growth. Aries and Scorpio feel Mars Retrograde more intensely than the rest of us. Find where those signs are in your own chart for clues on how this Retrograde can affect you.

THE OUTER PLANETS

Now, the Outer Planets stay in a Retrograde position for a long time. I honestly don't freak out on any of these Retrogrades, but we do seem to notice the days around when one of these planets moves backward or forward. Obviously, there are a few "rules" that correlate with each of the Outer Planets, but on the whole, they are not nearly as big a deal as when Mercury, Venus, and Mars move backward. Whew!

JUPITER

When the planet of luck and opportunity goes backward, it doesn't mean all of your good fortune goes out the window. But it *is* especially cool to calculate the days when Jupiter shifts Retrograde and then back again. We absolutely feel big Jupiter shifts during this time. So when the Planet of Expansion is moving backward, don't take a gamble or make a big push for prosperity. Go slow with taking risks or setting sail with new ventures. Use this Retrograde as a time to reevaluate and wait until there is forward momentum here. During Jupiter's Retrograde, issues of excessiveness and greed come up for reevaluation. Of course, it all depends on exactly where this occurs in your own chart, but when Jupiter is back on track, so are we. Sagittarians especially are tuned in to the movement of Jupiter.

SATURN

When Saturn is Retrograde we feel the need to restructure our plans and our goals. It's a reflective period that can be pretty damn constructive. Saturn loves to take its time and form a plan, and that's exactly what we end up doing during the backward motion of Saturn. This is also the planet of karma, so the five-month-long phase of this Retrograde has us admitting accountability and responsibility for recent actions and mistakes. It's always a good time to say "I'm sorry" during a Saturn Retrograde. Saturn rules business, so never, ever, *ever* start a new enterprise when this planet isn't in forward motion. But of course, it's awesome for tightening up plans and changing direction if need be. Capricorns feel this Retrograde the most.

URANUS

Uranus rules extreme and sudden change and when this planet is going backward it's not the best time for a big external transitional push. It is, however, a fantastic time for reawakenings and breakthroughs. Uranus is extreme and shocking and under its backward dance, the energies of Uranus are much less outwardly drastic and harrowing. We get in touch more easily with inner fluctuations and transformations. We crave freedom and some of us become more detached. Some of us rebel. But at its core, a Uranus Retrograde is an essential, internal shift that translates into reality when Uranus is forward once again. Aquarians are very in sync with this Retrograde.

NEPTUNE

When the foggy planet of dreams and fantasy turns inward, our spiritual life becomes astonishingly rich and incredibly colorful. This is a wonderful phase for deep contemplation and restorative reflection. Dream time becomes therapeutic and art and music offer beneficial rehabilitation into our psyche. Fantasy life is increased, but addictions and escapism turn excessive. Boundaries are blurred and self-sabotage is heightened. Paranoia increases, too, as does anxiety, for what is real and what is imagined? Meditation, yoga, and journaling are most helpful, and Pisces is always the most affected.

PLUTO

When Pluto shifts into retrograde it intensifies and deepens whatever house or planet it is activating in our chart. It helps us evolve and excavate. We discover our true motives and what's really holding us back. Pluto Retrograde reveals our shadow side, and our worst selves emerge when the Planet of Transformation moves backward. Deep psychological mutation occurs during this beautiful and lengthy Retrograde. And when Pluto moves direct, we are prepared for epic change with beautiful new clarity. Scorpios gain incredible insight when Pluto is Retrograde.

6

UNDERSTANDING THE ASPECTS

Once you have a relatively good understanding of the planets and the signs, it's time to learn how to put it all together. Enter the aspects. Aspects are simply the way in which the planets are communicating with one another. It's a bit like geometry, but please, don't let that turn you off! An aspect is simply when a planet is a certain number of degrees away from another planet or chart point. Computers now do all the hard work for us these days in terms of calculation (thank goodness!) and any chart software or website can usually tell you if, for example, your Pluto is forming a square or a trine to your Venus.

The angles in our charts uncover the dialogue between the planets. It discloses how the "actors" or planets get along . . . or how they clash. Good or easy aspects include *trines* and *sextiles*. The hard aspects are the *oppositions, squares,* and *inconjuncts. Conjunctions* are neutral depending on the two planets sitting next to each other. They can get along beautifully like Venus and Jupiter, or fight like cats and dogs, as with Mars and Saturn. Just remember that while the easy aspects feel like a nice, cool breeze, it's the arduous angles that push us and help us flourish—give us ambition galore!

THE MAJOR ASPECTS

CONJUNCTION

A conjunction occurs when two planets are very close to each other, usually in the same sign, but that's not always the case. Think of a conjunction as a blend of two planets, similar to a fusion restaurant. And just like classic fusion cuisine, it can be a phenomenal marriage or completely unappetizing. It all depends on the ingredients or planets that are stuck together.

Jupiter in a conjunction makes everything bigger: bigger temper (Mars), bigger creativity or addiction (Neptune), or a bigger appetite (Moon). Uranus can make planets act even nuttier or more unstable. Imagine Uranus with Venus—the planet of love. Or Mars, the planet of anger. Yikes! What about planets next to the Sun, which rules our identity. If you are born with Pluto, the planet of power conjunct your Sun, your identify is wrapped around prestige and control. Venus conjunct the Sun renders one incredibly charming and gracious, while Mars gives one an active and temperamental personality.

Stellium

If you have three or more planets conjunct one another, you have a stellium. When this occurs, the planet's meanings sort of weave together and they intensify whatever sign and house that stellium is in.

TRINE

Trines happen when two planets are around 120 degrees apart and usually occur within the same element. So if you have Uranus at 3 degrees Libra and then Mercury at 5 degrees Gemini, then voila! You have an Air trine. Got a late Aries Moon at 28 degrees? See another Fire planet near that same degree? Boom! You have a Fire trine.

Trines bring ease, luck, and cooperation. They feel great and each planet blends with the other harmoniously and effortlessly. Trines also make us lazy and boring because they don't require any extra work or effort.

If you are born with Mars trine Venus, you easily blend the sensual powers of love with passion. You're a natural when it comes to seduction and you don't even need to work on this. Lucky, sure, but while you may be great at snatching a date, your conversational skills may be lacking. Or perhaps you're great with the opposite sex, but you have hardly any friends. You've never had any problems attracting someone, so why work harder? You see, a trine can make you dull. Too many of these guys and life is too easy. Be sure to challenge yourself!

Got three planets all in a trine? Lucky you! You have what's called a *grand trine*. It's rare and auspicious and it can sure take you places. Just don't depend on that grand trine too much. If you never push yourself or struggle, you'll never be interesting.

ARIES, LEO, AND SAGITTARIUS form Fire trine or grand Fire trine

TAURUS, VIRGO, AND CAPRICORN form Earth trine or grand Earth trine

GEMINI, LIBRA, AND AQUARIUS form Air trine, or grand Air trine

CANCER, SCORPIO, AND PISCES form Water trine or grand Water trine

SEXTILE

A sextile isn't a strong aspect, but it's a good one nonetheless! It occurs when two planets are 60 degrees apart, and there is almost always an opportunity here. A sextile is usually between Fire and Air elements, or Earth and Water, and there is always a sign between the two planets in a sextile if you are trying to spot one. They're easy and get along rather well, but as with all the easy aspects, we can ignore them and miss out on their beneficial powers. Is your Mercury sextile Pluto? You have the opportunity as a healer and you can be powerfully persuasive; however, you still have to do the work and put forth the effort. Perhaps you're a high-powered lawyer by day, but your friends also come to you for your therapeutic tendencies at night. It's your sextile between those two planets that gives you these subtle abilities and it's up to you whether you decide to use it or not. Sextiles are like secret hidden abilities. Find yours and use them!

SQUARE

Now we are getting to the good stuff! The hard aspects are like the meat and potatoes of a natal chart. They'll build character and will sometimes even put hair on your chest.

A square happens when two planets form a 90-degree angle and are usually in the same quality. So you can have a Cardinal, Fixed, or Mutable square(s). These planets don't get along and they fight or "square off," so to speak. They struggle, and you'll struggle, too. It's an internal conflict and it's very, very frustrating.

We need tension in our lives to get us going, but too much and we get stressed. A square reveals what we can't stand within ourselves, and sometimes we even take it out on other people. These suckers are *tough* when we're young. Either we just get used to them as we age, or we learn to handle these tensions with a bit more ease.

Typically, there's one side of the square we prefer and the other side we inhibit. The strange thing is, we can switch sides a few times a day, or even every couple of years. For example, let's look at a square between Saturn and Uranus. Saturn follows rules while Uranus rebels. People who have this aspect vacillate between doing what's expected of them and then burning everything to the ground. They'll stick with being "reliable" and "good" and then all of a sudden throw caution to the wind and run away with the circus.

A better way to handle this is to give yourself some flexibility (Uranus) while being structured and regimented (Saturn). Learn to give attention to both sides of a square. Repress one planet and things can really get out of control.

ARIES squares off with Cancer and Capricorn
TAURUS squares Leo and Aquarius
GEMINI squares Virgo and Pisces
CANCER squares Aries and Libra
LEO squares Taurus and Scorpio
VIRGO squares Gemini and Sagittarius
LIBRA squares Cancer and Capricorn
SCORPIO squares Leo and Aquarius
SAGITTARIUS squares Virgo and Pisces
CAPRICORN squares Aries and Libra
AQUARIUS squares Taurus and Scorpio
PISCES squares Gemini and Sagittarius

OPPOSITIONS

Planets in opposition sit 180 degrees apart, directly across the other side of the Astrology Wheel from one another. If you have an opposition in your chart, then good on you! You get to see two sides of a scenario, but that isn't easy. Just like nodes, they highlight completely different viewpoints but can also balance one another out—if you work on it.

Oppositional aspects are at odds with one another. But unlike a square, where the energy is internal, an opposition almost always deals with another person. It'll give you great perspective when you learn to see both sides of the situation. Got Mars in Aries opposite your Moon in Libra? Perhaps you play Mars and get aggressive with the women (Moon) in your life. Sometimes that gets you nowhere, so then you flip the script. You might attract bullies (Mars) and play weepy and emotional (Moon). Remember you do one side of the teeter-totter and subconsciously pick out someone who can manifest or mirror that planet on the other side.

Let's say you have Venus in an opposition with Neptune. If you choose to play Venus, when you're in love, it always seems that the other player is always disappearing or ghosting you (Neptune); however, you might notice that when you reverse the roles and do Neptune, evasive and dreamy, the other player comes bounding back full of love. So, what to do? Well, a hard way to handle this is to always pick out the drug dealers and con artists, or waft around in a delusional dream when it comes to romance. You see what you'd be doing there? Always playing Venus and letting someone else play Neptune. Either way, it's easy to get disenchanted in relationships here. A smart way is to balance the two. Be ethereal and intangible while you are loving. Let the other person use both sides of the seesaw and you do the same. An opposition helps bring our conflicts to life. We learn to understand the side we choose and the side that we gifted to our opponent. Oppositions can absolutely assist us in seeing the big picture!

ARIES is opposite Libra
TAURUS is opposite Scorpio
GEMINI is opposite Sagittarius
CANCER is opposite Capricorn
LEO is opposite Aquarius
VIRGO is opposite Pisces

MINOR ASPECTS

INCONJUNCT AKA QUINCUNX

An inconjunct is when two planets are 150 degrees apart. Almost in opposition, but one sign over. Usually, they live in an entirely different modality and element. You can't balance them like oppositions, or use them like squares, either. It's a sneaky little angle and there's no way the planets can ever get along, so we learn to separate them.

Let's take the Moon and Mars, for example. Let's say you have them in an inconjunct and you've learned to isolate your feelings and emotions from your anger. You make adjustments and you never blend the two or even let them take turns. It's as if your aggressive side and your sensitive side exist in two entirely different worlds. Perhaps you get mad, but hardly ever feel guilty about it. Or maybe you feel so sensitive you can't tell other people when you are upset. You don't know how to blend the energies of both your emotions and your anger. It's a challenge for you and everyone around you.

Becoming aware of your own inconjuncts is key. Once you realize how you compartmentalize each planet, you can learn how to healthily express each planet involved.

ARIES is inconjunct with Virgo and Scorpio
TAURUS is inconjunct with Libra and Sagittarius
GEMINI is inconjunct with Scorpio and Capricorn
CANCER is inconjunct with Sagittarius and Aquarius
LEO is inconjunct with Capricorn and Pisces
VIRGO is inconjunct with Aquarius and Aries
LIBRA is inconjunct with Pisces and Taurus
SCORPIO is inconjunct with Aries and Gemini
SAGITTARIUS is inconjunct with Taurus and Cancer
CAPRICORN is inconjunct with Gemini and Leo
AQUARIUS is inconjunct with Cancer and Virgo
PISCES is inconjunct with Leo and Libra

SEMI-SQUARE

This is when two planets are 45 degrees apart from one another. Yep, you got it, it's half of the 90-degree square. A semi-square is uncomfortable but unlike a square aspect, it's way less apparent. This makes it much more difficult to understand. So if your Sun is semi-square Saturn, you're driven and hard on yourself but it comes across as almost trivial friction. In reality you may not work on yourself as would be the case if you had an actual square. Think of a semi-square as more of a hidden irritation. Once you are aware of it, you can absolutely use it to your advantage. Knowing where your semi-squares lie, you can push yourself much more assertively.

QUINTILE

This positive aspect occurs when two planets are around 72 degrees apart and usually corresponds to a hidden talent. In actuality they light up subtle conflicts between the two planets, but with this aspect, we are able to infuse this conflict with creativity and brilliant solutions.

SESQUIQUADRATE

A sesquiquadrate is difficult to pronounce and equally difficult to grasp when you have it. It happens when two planets are 135 degrees apart and it's similar to a square and semi-square. It highlights our tensions and strifes but since it's only a minor aspect, it's covert and hard to pin down. And just like with every tough aspect, it can absolutely reveal a great deal about our internal strength and how to use it.

Aspects are a challenging and complicated part of astrology, so don't beat yourself up if you feel a bit lost at first! Remember, the best way to learn is to study not just your own chart, but the charts of your friends and family, too. But it does take time. Before you know it, you'll start picking up on all of these crazy aspects, understanding how they play out, even when they are contradictory (and they often are). We humans are complex, paradoxical creatures. So naturally our cosmic blueprints are equally intricate and convoluted as well!

7

THE ASTEROIDS & CHIRON

Asteroids are the newbies of astrology. Most were not discovered until the 1800s, and then there's Chiron, which was only discovered in 1977. It's no coincidence that the asteroids, with the exception of Chiron, are all named after goddesses of mythology. The discovery of these suckers coincided with the brink of the women's movement during the turn of the twentieth century. Yep, these gals are the true feminists of the Zodiac. And you'd better believe we women love to sprinkle these babies around the chart wheel.

What Is Chiron?

Chiron is an asteroid that orbits our solar system somewhere between liberated Uranus and uptight Saturn. According to Greek mythology, Chiron was a centaur who happened to be a skilled healer and—a little ironically—could not heal himself from an arrow wound. Chiron's placement in your chart points to a core wound that may take a lifetime to heal. But don't despair, Chiron is also shaped like a key, reminding us that as we grapple with the pain, we can also open a path to inner peace.

These days there are hundreds and hundreds of asteroids, with more being discovered every day. In fact, according to NASA, the current number of asteroids to date is 958,512! Of course, this means it's relatively easy to fill up one's chart with chunks of rock, so let's keep things simple, shall we? There's really no reason to complicate the already complex world of astrology with excess comets, exoplanets, planetoids, satellites, etc. We will look at the main asteroids now commonly being used in Western astrology: Ceres (how you feel nurtured and provide love), Pallas (wisdom/intelligence), Juno (partnership needs), Vesta (sacred sexuality/hidden yearnings), and Chiron (hidden wounds).

So, why do we pay attention to *these* specific asteroids? Well, frankly, because we know they work. Even though they're subtle and the planets will always pack *way* more of a punch, it's no small thing to be born with Juno conjunct your Sun, or Ceres near your Ascendant. They'll give your basic Leo Venus or Virgo Moon way more zest and flavor.

The reason we know they work is we astrologers have had some time here. Not as much time with the easier-seen planets that we've been studying for thousands of years, but enough to know that Juno is linked to marriage and Pallas can reveal our intelligent and creative perceptions. We've studied them, and in the not-too-distant future, we'll have heaps of other asteroids that will work rather well, too.

CERES

Ceres was the first of the asteroids to be discovered, but in 2006 she was reclassified as a dwarf planet. She is named for the Roman goddess of agriculture, grain, and fertility and in a horoscope she is associated with motherhood and sustenance. She'll show what you need to feel nurtured and protected and in turn how you nurture right back. She'll also reveal your relationship with food, whether that be healthy or dysfunctional, along with the Sixth House and the Moon. FYI, I've seen many great chefs with a prominent Ceres placed in their chart.

Ceres, or Demeter in Greek mythology, is most famous for the loss of her daughter Persephone, who was abducted by Pluto and forced to live the rest of her days in the Underworld (until Zeus bargained with Hades to return Persephone to her mother for half the year, which explains the seasons). Ceres is mythologically well-known as the great mother, as she never stopped searching for her beloved daughter and never married. This is why she is associated with both Cancer and Virgo: Cancer for motherhood, and Virgo for food.

Ceres strongly placed in a chart indicates a mothering figure in a woman, and suggests robust feminist qualities in a man. A strong Ceres occurs when you have the nurturing asteroid on your Ascendant, Midheaven, or a personal planet, like the Sun, Moon, Mercury, or Venus. Astrologer Kim Falconer has noted that Phil Donahue, the groundbreaking talk show host who spoke out for women's rights, has Ceres conjunct his Moon and Venus. It's interesting too to note that in *The New American Ephemeris* (a book or table that denotes future planetary positions), Ceres is now included among the eight planets along with the Sun and the Moon. It is the only "asteroid" to be acknowledged thus far. The symbol for Ceres is the sickle, as she also rules over wheat and grain production.

CERES IN THE FIRST HOUSE

Ceres in the First reveals a truly nurturing individual. It's part of your persona and you may find people lean on you quite a bit. You're like their mom.

CERES IN THE SECOND HOUSE

You feel nurtured and protected with money. This is a wonderful place for Ceres. She's very protective of your resources here in the second. Lucky you!

CERES IN THE THIRD HOUSE

You make a wonderful counselor or salesperson, as your words are protective and nurturing. The Third House also rules siblings, so perhaps you had to act as their mother, or they for you.

CERES IN THE FOURTH HOUSE

The motherhood asteroid in the house of home and family makes you a natural when it comes to caretaking. You're instinctual with knowing what people need, and may have an uncanny knack for real estate, too.

CERES IN THE FIFTH HOUSE

You nourish yourself with art and perhaps a good bit of risk-taking. You're good with kids, and if you have them, your firstborn is probably quite mothering and also a great little chef.

CERES IN THE SIXTH HOUSE

You love being useful, and your daily routine is something that makes you feel serene and secure. This also indicates someone who could be in the medical field or a nutritionist. You're probably a great cook.

CERES IN THE SEVENTH HOUSE

Similar to having the Moon or Cancer in this house, you either mother in a relationship or you like to be mothered. It'll be easier if you take turn with this placement. Too much one-sided unconditional love and you may find yourself feeling resentful.

CERES IN THE EIGHTH HOUSE

Ceres in the house of sex and others people's money? Yes, you feel nurtured by both, you sexy beast! The Eighth House is also the house of death and rebirth, so crises makes you comfortable.

CERES IN THE NINTH HOUSE

You can nourish yourself by jet-setting off to far and distant lands, or at the very least amassing a great amount of knowledge about them. You also have a deep hunger for learning, and probably do a great deal of reading.

CERES IN THE TENTH HOUSE

You may make a living out of nutrition or caretaking, especially if Ceres is conjunct the Midheaven. Work is where you feel most nurtured and you also make a great manager. People feel comfortable with you!

CERES IN THE ELEVENTH HOUSE

You mother your friends! You also throw the best parties and make one hell of a host. You have a great talent for getting people together in your community and could really make a social impact, if you wanted to.

CERES IN THE TWELFTH HOUSE

The Twelfth House is where things are hidden. Ceres here might indicate that your own needs weren't quite met when you were younger. It also means that you can surprise yourself and others with your own nurturing abilities. It's there, all right, just quietly tucked away.

PALLAS

Pallas Athena was the second asteroid to be discovered and is named for the goddess of war and creative intelligence. As the story goes, she was born from Jupiter's head, springing forth fully dressed in golden armor and ready for battle. She wants peace like Libra, but unlike Libra, she loves to fight. Pallas is just and fair and relies heavily on her creative intellect and pattern recognition for her many successful victories.

In a natal chart, Pallas represents wisdom, healing, and strategic intuition. She's courageous and innovative and signifies how we can be creative, too. Pallas is also a phenomenal problem-solver. She'll show us how we can figure things out with logic and reasoning but with an artistic twist. Perception is closely tied with this asteroid.

Pallas was famously celibate, preferring to reserve her strength for battle instead of wasting it on men or women. In this regard some look at Pallas with slight androgynistic tones. David Bowie had Pallas Athena in his first house; in fact, he even wrote a song about it!

Pallas is genius and when strongly placed in a woman's chart can designate an alpha female type. Aviation pioneer Amelia Earhart also had Pallas conjunct her Ascendant. There was no way she couldn't do what a man could! Pallas Athena is linked with Libra, Leo, and Aquarius. Her symbol looks like Venus with a diamond-shaped shield, and some astrologers also associate her with an owl and a snake, in honor of her cunning and insightful nature.

PALLAS IN THE FIRST HOUSE

You're clever and, like a wise old owl, can see in many different directions all at once. You probably don't play into the typical male/female stereotypes, which renders you rather unconventional.

PALLAS IN THE SECOND HOUSE

You are a genius when it comes to finances. You have mad skills at manifesting your resources and are quite astute with means of procuring your security.

PALLAS IN THE THIRD HOUSE

You can win wars with your words. Whether that be diplomatically or aggressively depends on other factors in your chart. But you are always deeply intuitive and highly intelligent.

PALLAS IN THE FOURTH HOUSE

You're a natural problem-solver with strong psychic abilities and gut instincts. You're emotionally intelligent and probably prefer living in a highly creative environment.

PALLAS IN THE FIFTH HOUSE

You're highly gifted when it comes to the arts. You see patterns and forms unlike most people and are blessed with strong creative impulses.

PALLAS IN THE SIXTH HOUSE

Pallas here can denote one who excels in crafting a well-oiled system or routine. This can also translate to health. You have strong reasoning powers and even stronger healing abilities. Especially with food.

PALLAS IN THE SEVENTH HOUSE

You're creative when it comes to relationships. You're no pushover, but you probably have some in-depth insight into the strategy of love and war. Honestly, it's all the same to you.

PALLAS IN THE EIGHTH HOUSE

You're a natural healer and maybe even a great therapist. You also possess admirable insight into shared business and money matters.

PALLAS IN THE NINTH HOUSE

You're a gifted teacher. You grasp the big picture, and can see patterns in everything. But the big thing when Pallas is in the Ninth is your ability to fight for the truth. And you do every time.

PALLAS IN THE TENTH HOUSE

You have epic career strategy and insight galore. It gets you to the top and keeps you there for the long haul.

PALLAS IN THE ELEVENTH HOUSE

You have strong visions and plans for your community and friend group. This is a wonderful place for Pallas, as she is the fighter of social justice here. You make a wonderful activist.

PALLAS IN THE TWELFTH HOUSE

You are of spiritual service and healing to those in need. You fight for the underdog, too. This of course makes you a natural healer and counselor. Meditation is therapeutic for you.

JUNO

Juno is the wife of Jupiter and the goddess of marriage and maternity. She was the queen of Mount Olympus and portrayed the extremely faithful and ever-loyal wife. She presided over all marriages and was an excellent matchmaker. The month of June is named after her and is still the most popular month for many brides-to-be. Juno is also associated with menstrual and reproductive cycles, as it was hers that she used to calculate time.

As beautiful and truthful as Juno was, her husband, Jupiter, had a terrible wandering eye. He had many lovers and quite a few offspring, too. This enraged Juno—rightfully so—and spawned a long and arduous partnership. This is why in a chart, Juno represents commitment, marriage, *and* betrayal. She is angered by any inequality and strives for balance in all forms of partnerships. Usually in the romantic sense, but we can absolutely check Juno in our charts for business and platonic relationships, too.

Juno is associated with Libra and Scorpio and her glyph looks like Venus with a star on the top for her royalty. Juno craves a healthy, sexual, and fair union. She'll show us what we need in a partner and what we get . . . whether we like it or not.

JUNO IN THE FIRST HOUSE

Your identity is closely intertwined with being in a partnership. Whether that be marriage, deep friendship, work spouse, bromance, or a business collaboration, it really is part of your persona.

JUNO IN THE SECOND HOUSE

You need a marriage that is stable, secure, and grounded. A good partnership is something also that helps ramp up your self-esteem.

JUNO IN THE THIRD HOUSE

A talkative, communicative partner suits you best. Variety is also equally imperative, and it's easy for you to get bored. You crave an active, intelligent union.

JUNO IN THE FOURTH HOUSE

You and your sweetie could possibly spend weeks at home and never leave and you're just fine with that. Watch for getting too cocooned here.

JUNO IN THE FIFTH HOUSE

Relationships unlock your creativity and perhaps you even attract a creative type yourself. If you choose to have kids, they can be very beneficial to your marriage. Of course, you need to look at the chart as a whole, too. There could be other factors that don't jibe with the whole kid thing.

JUNO IN THE SIXTH HOUSE

You can actually work with your partner, which is no small feat! Or perhaps you're simply married to your work. In any case, you are a workaholic and you need a partner who understands that.

JUNO IN THE SEVENTH HOUSE

Marriage or solid partnerships are a central core theme when you have Juno in the Seventh. It's part of your growth and evolution. This is a great place for Juno!

JUNO IN THE EIGHTH HOUSE

You yearn for someone intense and highly sexed and don't be surprised if your relationship undergoes many different transformations, too. Marriage can also assist you in your own personal transformations.

JUNO IN THE NINTH HOUSE

You need a partner who expands your mind and you also might attract people from other countries. Someone who has the same basic belief systems is crucial, too.

JUNO IN THE TENTH HOUSE

You perhaps gain social status from your partnership or you can be married to your career. Perhaps you even met your partner on the job. Juno in the Tenth denotes a strong correlation between marriage and your work path.

JUNO IN THE ELEVENTH HOUSE

Friendship is crucial with this placement. You need to marry your friend, or you just won't be fulfilled. You also might attract someone who is somewhat of a social activist, too.

JUNO IN THE TWELFTH HOUSE

There is a karmic quality to this placement of Juno. It's also all too easy to be blindsided when Juno is here, but as always, that depends on lots of other things, too. Relationships to you take on a mystical, spiritual union.

VESTA

Named for the Roman goddess of the hearth and the sacred flame, Vesta was the fourth asteroid, discovered in 1807. She is associated with the Vestal Virgins who protected her fire and kept it burning continuously in the center of town. In fact, every home in Rome had a constant sacred flame thought to bless and protect it. It was never allowed to extinguish and even when you moved, your eternal flame moved right along, too.

The Vestal Virgins were noble Roman girls who were chosen around age six and took a vow of celibacy for thirty years. Their duties were to keep the holy and blessed fire going along with various rituals and offerings. It was an honor to be chosen as one of Vesta's servants, but there were dire consequences if Vesta's flame was ever suffocated, and death awaited anyone who broke their vow of chastity, though, of course, after thirty years they were free to marry as they wished.

Vesta is the brightest asteroid and in a horoscope chart she is associated with both Scorpio and Virgo. She represents sex, dedication, and commitment. But Vesta can show where you can sacrifice, too. Vesta reveals a strong feminine side in a woman, and also reveals a man who may be surrounded by women in a man's chart. Cancerian Richard Branson has Vesta in his career house. He began his Virgin Group in the 1970s and it later expanded to Virgin Records and Virgin Atlantic Airlines. Fascinating, eh?

VESTA IN THE FIRST HOUSE

You're dedicated to self-discovery and personal goals. However, be careful not to sacrifice other people and relationships in your quest to know thyself. You thrive on solo retreats.

VESTA IN THE SECOND HOUSE

You're committed to security and material things. You're a hard worker, too, though you know what it means to sacrifice in lieu of nice things.

VESTA IN THE THIRD HOUSE

You are devoted to communication and learning on all levels. However, you might need to withdraw and give that big brain of yours a rest every now and then. You're a phenomenal multi-tasker; just make sure you don't have too many irons in the fire.

VESTA IN THE FOURTH HOUSE

You are loyal and faithful to your family and home but there is a possibility of too many obligations and responsibilities that can leave you drained. Remember: dedication to your family does not always require self-sacrifice!

VESTA IN THE FIFTH HOUSE

You're devoted to your children if you have them, and crazy dedicated to self-expression and creativity. In fact, you may be well known for your artistic skills when Vesta is here, devoted to your craft.

VESTA IN THE SIXTH HOUSE

You're dedicated to health and nutrition and a very effective schedule. You like things to run smoothly, which is admirable, but watch out for overdoing it on the perfection front. You want things too exact at times.

VESTA IN THE SEVENTH HOUSE

You are loyal and dedicated to relationships. Whether romantic, platonic, or business, you put a lot of energy into the other people in your life. Too much, though, and you'll sacrifice yourself for someone else. Just remember to incorporate some balance here.

VESTA IN THE EIGHTH HOUSE

Vesta in the House of Sex does not make one celibate, although watch what you can surrender here. Vesta in the Eighth denotes a lot of energy poured into deep intermeshing with another. Sex, psychology, and the occult are big commitments here.

VESTA IN THE NINTH HOUSE

You are dedicated to higher learning and travel but watch for obsessiveness with politics and/or religion. You can become unbearable if that's the case.

VESTA IN THE TENTH HOUSE

You are insanely committed to your career and public reputation, and usually quite successful. But remember, Vesta also requires compensation, so be aware of what you are giving up in order to achieve your high status.

VESTA IN THE ELEVENTH HOUSE

Faithful, devoted, and unwavering, you make a wonderful friend. You thrive in groups, too, but remember to focus on your own hopes and wishes. No need to sacrifice what you want for everyone else.

VESTA IN THE TWELFTH HOUSE

Vesta in the mystical Twelfth House denotes one who may sacrifice a great deal for spiritual pursuits. You're a hermit at times, with strong, healing psychic abilities and the gift to transverse both the conscious and unconscious realms. You're very, very compassionate.

CHIRON

Now, Chiron is a bit of a black sheep in this group. Discovered by Charles Kowal in 1977 and named Asteroid 2060, Chiron was later classified as a comet, but is now considered a dwarf planet or planetoid, with a tail. In Roman mythology, Chiron is a centaur. Born from Saturn and Philyra, he was a teacher, philosopher, and healer who could not heal himself from an enchanted arrow wound.

The orbit of Chiron is strange and erratic, not unlike that of a maverick, and that's exactly what he represents in your chart. Some say he is the Wounded Healer, but astrologer Zane Stein (who I think has written the best book on Chiron yet) suggests that Chiron is a key for opening the doorway between Saturn and Uranus. Saturn wants to uphold tradition while Uranus chooses to break down walls. So, what is it that Chiron is unlocking for you? He neither depends on nor overthrows establishment. Chiron doesn't fit in . . . happily so, and he'll show you where and how you're a maverick, too!

CHIRON IN THE FIRST HOUSE

You have "loner" written all over you, but you don't really care. You know what people do to fit in and you'd rather not participate in any fake, superficial crap, thank you very much. But you're a wonderful teacher and healer, always fighting for the underdog.

CHIRON IN THE SECOND HOUSE

You learn a great deal about your values and resources with Chiron here. You're also incredible with money once you learn to bridge the idea that you don't have to suffer to make money, or rebel to throw it all away.

CHIRON IN THE THIRD HOUSE

Yes, your words are healing, though you might spout some crazy ideas! You have a distinct style of talking and writing that drives folks to you. Remember, people are listening.

CHIRON IN THE FOURTH HOUSE

You're exceptionally keen to get along with all types of people from all walks of life. Everyone is family to you, but you have a deep fear of letting people down. Don't let that stop you or hold you back.

CHIRON IN THE FIFTH HOUSE

You are a natural-born creative who doesn't need inspiration from anyone else. Your self-expression is yours and yours only. If you have children, your kids can teach you a lot about life.

CHIRON IN THE SIXTH HOUSE

You have a strong sense of duty and responsibility but only when it comes from within. Outside pressure never quite works on you. You often rebel until you realize being of service is the key to unlocking your own spiritual quest.

CHIRON IN THE SEVENTH HOUSE

Relationships are important to you, but you're not so great at cooperation. You have an oddball way of relating but partnerships are healing for you.

CHIRON IN THE EIGHTH HOUSE

You're incredibly powerful and deeply sexual . . . though your tastes may run a bit strange. You seek out the bizarre and unusual and are able to channel it into amazing healing abilities. Your control and self-discipline are, frankly, extraordinary.

CHIRON IN THE NINTH HOUSE

Your strong perceptions and ability to intuit the facts are your true talent. You're super outspoken and make a fun teacher and are also great at debate. Conversation is your talent and you've learned the hard way.

CHIRON IN THE TENTH HOUSE

Here's a hard worker who makes it to the top in a nontraditional way. You don't seem to fit in with the rest of your colleagues or perhaps you do things a bit differently, but it hasn't stopped you yet. Basically your maverick abilities are on full display here.

CHIRON IN THE ELEVENTH HOUSE

You have a peculiar set of friends you have specifically crafted who are probably very different from any friends you had when you were younger. Your ideals are strong and your friend group needs to match that.

CHIRON IN THE TWELFTH HOUSE

You want to learn everything and anything around any hidden motives or truth behind people, situations, and scenarios. You also have supernatural powers and you know there is much more to life that lies behind the veil.

Remember that most of your chart's important themes can be found working through the planets—and Sun and Moon. However, those asteroids out there can be helpful in pointing out secondary themes and giving you a deeper, richer perspective.

8

HARNESS YOUR MOON WISDOM

The power of the Moon has enchanted people for as long as anyone can remember. A fascinating and mysterious luminary, she has inspired many to craft beautiful and terrifying tales of myths and lunar lore. Werewolves and humans alike come alive when the Moon is full. Lunatics, too. Waxing and waning, the Moon is a constant reminder of change and this natural satellite casts a spell over us with her many mystical faces and rhythms.

Our Moon is a seductive presence, but she's also crucial for Earth's livability. She's the closest celestial body to Earth and she dictates seasons and assists in controlling a stable environment for all living creatures. As any *Farmer's Almanac* will tell you, lunar cycles play a big role in farming and gardening. It'll tell you the best time to plant, weed, and harvest. The Moon absolutely corresponds to growth periods, and it's not just for vegetation! She's also linked to fertility for many species, and even plays a role in syncing women with their menstrual cycles. The Moon also controls the tides and since we humans are more than 60 percent water, it only seems logical that we'd be impacted, too.

Fun Fact

Did you know that our Moon is actually egg shaped? Thanks to our planet's gravitational pull, the large part is what faces Earth, which is what gives us the impression that the Moon is full and swollen.

It takes the Moon about twenty-eight days to orbit Earth and the Astrology Wheel, and it changes signs every two-and-a-half days. The Moon governs the general public and we are all affected when she's in a void period (when the Moon has made its last aspect to another planet before it changes signs) or when she's in a temporary yet difficult configuration with another planet. Learning to work with the Moon and her phases will definitely give you an advantage. It's phenomenal for goal setting and manifestation. You can't go wrong when you plan your life by the Moon.

As she passes through her many different phases, the Moon creates a rhythmic sort of dance, one that we can easily tap into and access. This lunar dance is ever-changing and it affects our moods, emotions, even our actions and attitudes. When we access the Moon's rhythmic nature, we can learn to go with her flow. Bask in opportune times, instead of fighting an uphill battle. How we respond to this lunar pattern is of course personal for each of us. But the general themes are the same. The relationship that the Moon has within our own chart is equally important. Her predictable monthly dance around our natal chart gives us an extra advantage. Tracking this is how we can tap into that lunar energy.

What's the difference among a New & Full Moon & an Eclipse?

So glad you asked! Thanks to Earth's wobbly orbit, New and Full Moon phases aren't always exactly aligned with the Sun and the Moon. A meticulous lineup is needed for an Eclipse to actually occur. There are three different types, too—Total, Partial, and Annular. They occur somewhere between two and five times a year. A Total Eclipse is the most rare.

PHASES

NEW MOON

Look up at night during a New Moon and you'll see nothing but dark skies. She's sandwiched between the Earth and the Sun, so we can't see her. This marks the beginning of the Lunar Cycle. Here the Moon is associated with fresh new beginnings and that's exactly what you'll crave during this phase. When we have a New Moon, the Sun and the Moon are conjunct in the sky and in your chart. We're more introspective and impressionable, and we prefer to be left alone and recharge. The New Moon phase is the best time to hit the reset button and start anew. Think of this as your clean slate. It's an amazing phase to plant new seeds and we are doing that metaphorically as well. New ideas, projects, mind-sets, and routines spring to mind. We begin to ruminate on our next venture.

Every twenty-eight days or so we have a New Moon and it's always activating our chart somewhere. For example, every Gemini season (May 21 to June 20) there is a New Moon in Gemini, and a Full Moon in Sagittarius every month of every year. Just like every Scorpio Season (October 23 to November 22) there is a New Moon in Scorpio and a Full Moon in Taurus. Every year. This knowledge can be a great advantage! When you start accessing lunar phases, you'll learn how to key in to when and how you create your own personal new cycles. You may even observe some eerie patterns: For example, each August you always seem to start a new home project—it may seem like chance, but perhaps you have Leo in your House of Home and Family. When that New Moon activates in Leo (July 22 to August 22), your desire to bring new energy to your home does, too. Or maybe every January you seem to be crazy focused on your finances, since Capricorn resides in your Money House. The New Moon phase is key to understanding how each month unfolds for your own personal growth and evolution.

Strategic planning will give you the best results here. Calculate when the next New Moon will activate your own chart. You can use this info to manifest new projects or launch a fresh beginning.

BORN DURING A NEW MOON? You're a visionary leader with strong instincts. You're spontaneous and prefer living in the moment.

WAXING CRESCENT MOON

When the Moon begins to move closer to the Sun you'll see that a tiny silver sliver starts to emerge, which we call the Crescent Phase. The Moon is beginning to wax, or grow, and thus your new idea or seed that was planted during the New Moon starts to see some expansion as well. If the New Moon is a seed, the Crescent phase is a seedling. It's usually a time of some struggle as we are forced to make decisions and restructure plans, but it is an exciting phase of new possibilities. Still, there is resistance here, too. Bringing life to new ideas requires release of old patterns and habits. Ones that need redirection and clarity. This is the phase where we gather necessary materials and tap into our ambition. The Crescent Phase wants us to clarify our plans and we usually make big breakthroughs here.

This is a time to set intentions. Get out your journal and write down five goals and five hopes. Look for alignment. Begin to plan the groundwork to achieve them.

BORN DURING A CRESCENT MOON? You're creative and rebellious, but vulnerable, too. You have a childlike, wide-eyed wonderment and feel like anything's possible.

New Moon, New You

At each New Moon, take some time to reflect.
This is a time to pause and clean your slate.

ARIES

What are some of your limiting beliefs? What are generational patterns that appear in your life that aren't serving you anymore? Write them down.

TAURUS

What makes you feel most grounded? Zero in on the factors that are most stabilizing in your life. Are you doing those things, and why or why not?

GEMINI

What words speak to your heart? Make a list. Ponder why they are there and how they show up—or don't—in your current life.

CANCER

When was the last time you had a good cathartic cry? How do you give yourself space and time to feel those more raw emotions?

LEO

How do you express yourself on a creative basis? Learning new ways to activate your playful and artistic nature is good for the soul and your self-esteem!

VIRGO

How is your current health routine nourishing you? What things can you do to recommit to your body's wellness?

LIBRA

Where do you feel imbalanced in your life? How can you create more equality and harmony in your relationships and in your independence?

SCORPIO

How satisfied are you sexually? What are some ways you can honor your body and embrace your sensual side?

SAGITTARIUS

What sets your soul on fire? This is a simple question, and yet, oh so powerful!

CAPRICORN

What aspects of your daily routine serve you? What aspects do you need to let go?

AQUARIUS

What is a big dream that scares you? What holds you back? What pushes you forward?

PISCES

How are you in service to others? What are some healthy ways that you can give more?

FIRST QUARTER MOON

This action phase begins around a week after the New Moon. This occurs when the Moon forms a 90-degree angle or square to the Sun—exactly one quarter of the way through her cycle, hence the name. This is a time when hard work is required to see our new intentions prosper. We've cultivated our new plan, brainstormed, meditated, and even made some headway. With the First Quarter phase, we really begin to make some progress. True, there may be some friction, but in general this is the point of no return. Which means we're really sharpening our skills and fixed on the outcome.

This is also the phase when we're presented with a crisis. Let's say your New Moon intention was to dramatically cut back on sugar. The First Quarter phase would be the day you happened to walk by a beautiful window display full of mouthwatering doughnuts. It's a mini crisis and your confidence is shaken. This is the time when you might hit some resistance to those intentions you wrote down during the Crescent Phase. Be flexible, be ready to pivot, but also be committed to work. If you remain open and flexible as challenges rear their ugly heads, the First Quarter phase will strengthen your resolve. Just remember to keep your eye on the prize and don't waver.

BORN DURING A FIRST QUARTER MOON? You thrive in conflict situations. You have a strong warrior energy and perhaps your parents didn't get along at the time you were born.

WAXING GIBBOUS MOON

Big and bright, the Gibbous Moon can be seen during the day—even when the Sun is out! She is coming into fruition and so are our intentions for the month. As the Moon nears its fullness, she invites us to refine and edit our goals. The Gibbous Waxing Moon makes us analytical and focused. Details are scrutinized and connections are made and prosper well under this phase as well. Think of this phase as the time to gather last resources. To bring in the people who can help you accomplish your goal, or who can cheer you to the finish line. The work hasn't stopped, but rather than encountering obstacles that you need to push through, focus more on adjusting and refining. This is both a confident and anxious time, but that finish line is in sight!

BORN DURING A GIBBOUS MOON? You're highly intelligent and have a great gathering of information. You question absolutely everything, too.

FULL MOON

The Full Moon is a time for epiphanies and extreme clarity. The two-week mark after the New Moon is the climax of our new intentions. It's like our cosmic harvest and Full Moons are always eye-opening. It's a powerfully potent time when the Sun and the Moon are in an opposition in the sky and in your chart.

Similar to the New Moon, a Full Moon is a seasonal event that occurs in the same sign roughly every time that year. It's actually quite cool to sync these up in relation to the New Moons that occur six months prior. Again, let's take the sign of Gemini, for example. Every year between May 21 and June 20 there is a Gemini New Moon. This means six months later there is always a Gemini Full Moon during Sagittarius season, November 22 to December 21. Every year. Pay attention to the themes that begin around each New Moon. There are always spooky similarities that feel like an ending or illumination six months later. A New Moon is a beginning; a Full Moon is more of a completion. Some of your hard work is coming into fruition. Make sure you are mentally prepared to receive new opportunities.

BORN UNDER A FULL MOON? You've got a *big* personality and are a natural performer. You're an idealist, too. You're drawn to relationships that play out in your own Sun and Moon opposition.

DISSEMINATING MOON

The Disseminating Moon appears around four days after the Full Moon and marks the waning period. This is the contemplation phase of the Lunar Cycle. We are fully conscious of whatever was exposed during the Full Moon, and now that we understand her message and revelations, it's time to share our experiences and move forward. There is an evangelical quality to this phase. We're excited about whatever we've learned and we're ready to tell the world. But be careful! This can be a gossipy time, too. Perhaps the Full Moon revealed a secret—remember to take care with whatever new information has come to light. This phase is all about what you do with your newfound enlightenment. The decision to pass it on or let it go is up to you.

This is also a time to be generous. You've just received some abundance from the Universe, so use this time to give back—buy a friend a coffee, treat someone to dinner, or make a donation to a favorite cause.

BORN DURING A DISSEMINATING MOON? You're an enthusiast and make a superb teacher and philosopher.

LAST QUARTER MOON

This occurs a full week after the Full Moon and marks a time when the Sun and Moon are square, once again. The First Quarter Moon marked a crisis of action; the Last Quarter Moon denotes a crisis of consciousness. It's a conflicting time where we must decide what to let go and what to keep as we move forward. We crave redirection and change. It's the only way to move onward with our new knowledge and awareness. The Last Quarter Moon is ideal for release: letting go of ideas, grudges, resentment, or plans. This is the phase to literally clean the house. The Moon's progress toward the Full Moon is accumulative. Now that it is passed, it's time to shed what we no longer need, making way for the dark of the Moon and the next phase. This is a great time to do a cleanse to be ready to set new intentions. Get rid of clutter and break a sweat.

BORN DURING A LAST QUARTER MOON? You're a natural activist and tend to go to more extremes than most people. You're also highly opinionated and very clever.

BALSAMIC MOON

This is the last phase of the Lunar Cycle and the best time for absolute surrender. The Balsamic Moon is also known as the dark moon phase. It's a fabulous period to go inward and recharge. We're not quite ready to start fresh or make new plans for the New Moon, but we are highly subjective and intuitive during this phase. In fact, the stiller and quieter you can become during this period, the more information you'll absorb. It's more like magical insight, not a full-blown revel. Meditation is fabulous during this phase. Unlike the Full Moon, this is definitely *not* a party Moon, and you'll want to tend carefully to your body and spirit in these last days of the cycle. You might be itching to set new intentions but wait for that New Moon energy. Right now just let yourself be. Rest. Wait. Be patient.

☽ **BORN DURING A BALSAMIC MOON?** You're a loner type, and perhaps very spiritual, too. This is a karmic life for you and you encounter people and situations that will assist you in resolving your past.

VOID MOON

Technically the Void Moon period is not a traditional Moon phase, but that doesn't make it any less important! In fact, it's a crazy absurd time when the Moon isn't making any aspects at all! It's a strange and aimless time that may be just several seconds or up to two days. It begins when the next moon fulfills the last main aspect with another planet before proceeding to the next sign when it doesn't have another aspect with a planet. Do you ever get out of bed with a lot of energy, mentally motivated to tick off all the items on your to-do list, and instead end up binge-watching an entire series on your favorite streaming service? When it's a Void Moon, you might as well take a seat, because there isn't much to do but wait. This cosmic event is a lot like a computer update: you might have a lot to get done, but until it's finished, you are going to be twiddling your thumbs. It's usually only a few hours, though, so y'all won't be adrift for long.

☽ **BORN DURING A VOID MOON?** You're unpredictable and it's probably hard for you to be grounded at times. Of course this makes you super creative and musically gifted as well.

9

TIPS & TRICKS TO TAKE YOUR PRACTICE FURTHER

Learning to work with the stars is not only fun, but therapeutic! By getting insight into who you are and what makes you tick, you become better at caring for yourself. In times of stress and anxiety—all too common these days—learning to connect with your basic desires is the first step to preserving your sanity.

Self-care is personal. After all, it's about caring for your one, perfect, unique self! It's only natural that what satisfies one person might stress out someone else. And that's okay. The goal is to figure out what works for *you*. Tapping into your emotional needs is not only key to increase your happiness, it'll also help you achieve your goals and uncover your true purpose in life. You can't possibly thrive in life until your basic needs are met, right? And we're not talking about shelter, food, and water, either.

So how, exactly, can we uncover what it is that can make us all breathe a little deeper and drift into a place of peace and tranquility? Well, in astrology, when we're talking about self-care we're talking about your Moon! The Moon reveals to us what we need and what fulfills us emotionally, the sign that your Moon is in will show the style of your needs, and the house placement also holds clues to the way your needs play out in the world.

MOON IN THE FIRST HOUSE OR ARIES

You need your needs met *fast*. Up and at 'em when your Moon is here! Procrastination and lounging around on the couch will only make you feel worse. So get up and get moving. A run is great, but really any form of exercise will do.

MOON IN THE SECOND HOUSE OR TAURUS

Turn off that phone and turn on Netflix. Close the windows and make a giant bowl of something with at least thirty-six grams of carbs. No one, and I mean *no one* can relax as well as a Taurus Moon. No guilt!

MOON IN THE THIRD HOUSE OR GEMINI

Get to talking! The last place you'd like to be is home, but if you must, burn a hole in that phone plan. Any form of socializing will recharge you in a heartbeat, but if you're extra blue, journaling may be the best solution.

MOON IN THE FOURTH HOUSE OR CANCER

Being around family or your friend tribe makes you feel needed and comfy, but if that means leaving the house, you may just need to veg. A good long soak in the tub will do, and looking at old photos makes you feel serene.

MOON IN THE FIFTH HOUSE OR LEO

Shopping nourishes your soul, but let's not blow your budget here! Simply going through your closet and trying everything on grounds you. Give yourself a makeover and a manicure, too. I have known a few Leo Moons who feel instantly rejuvenated with a quick swipe of a bold, red lip.

MOON IN THE SIXTH HOUSE OR VIRGO

Organizing and rearranging your space de-stresses your cluttered and anxious mind. Meditation is always a good idea for Virgo Moons, and fresh clean sheets scented with lavender or patchouli invoke much-needed sleep for your overactive brain.

MOON IN THE SEVENTH HOUSE OR LIBRA

You're a beauty product hussy, so delve into all those unopened tubes and potions lurking around your bathroom for the ultimate in self-care ritual. If you're too blue for primping, balancing yourself will feel oh-so-restorative. Yoga and a fresh bouquet of flowers do you nicely.

MOON IN THE EIGHTH HOUSE OR SCORPIO

Sex is your emotional go-to. To you, it's like brushing your teeth and eating healthy, nourishing food. It should be a weekly or daily part of your routine, but if you're extra down, nothing transforms you better than some isolation with your music.

MOON IN THE NINTH HOUSE OR SAGITTARIUS

Okay, you Moons don't stay down for long, but when you need a pick-me-up, a good book, vacation, or travel documentary is your best bet. You're not a homebody, so even a good walk around the neighborhood lifts your spirits. Bonus points if you throw in a good podcast.

MOON IN THE TENTH HOUSE OR CAPRICORN

It's easy for you to overwork yourself, so it's imperative that you actually schedule in some downtime every once in a while. A good idea is a daily gratitude journal to get you in a more positive frame of mind. And exercise gets you all glowy, too!

MOON IN THE ELEVENTH HOUSE OR AQUARIUS

You feel invigorated in groups of people. So taking a class or enrolling in any sort of health club gives you all the feels. Of course, no other Moon needs space as much as you, so be aware when you just need a break. New, techy stuff refreshes your soul, too.

MOON IN THE TWELFTH HOUSE OR PISCES

Spirituality refuels you and there are so many ways to access that euphoric feeling. Meditation and yoga are great ways to do it, but stay away from the wine when feeling blue. It will only deepen your depression in the long term. Pedicures and foot massages are equally calming.

If you're down in the dumps, look to the house *and* the sign that your Moon is in. It'll reveal what makes you feel comfy and show you the best way to make it all better. Just like Mom used to. Of course, there are also other ways to nourish your soul with ritual and intention. Once you have a basic idea of what your own needs and wants are, creating a spiritual sort of ceremony is a fabulous way to boost manifestation!

CREATING MEANINGFUL RITUALS

We don't have to get too woo-woo here, but astrological rituals can be an awesome way to bring in more oomph to your Sun sign, Rising sign, or really any planet that you are trying to get a bit more connected with. Putting energy and intention into the vibe you'd like to bring into your day can be as simple as lighting a candle or incense or just taking a bath. I find myself doing these magical little rituals when I need to up my energy, my focus, or light up specific planets in my chart. Activating a particular planet by sign and by house helps you bring focus and attention to what you feel like you are lacking. For example, if I'm feeling low, I'll infuse a bit of candle magic and written intentions to tap into my natal Mars. If I need to bring in a more empathetic vibe, I'll access my Moon with a candlelit bath.

You can make these little ceremonies as complicated or as easy as you'd like! All the signs correspond to certain colors and certain oils or fragrances, so that's often an easy place to start. Of course, use your intuition if something feels off or doesn't seem to fit for you. And remember, this is not rocket science, so you don't need to get stressed or all out of sorts with this. This is simply a loose guide to help assist you in your own journey of activating your chart. When you incorporate daily rituals into your life, it'll give you deeper meaning and connection not only with your day-to-day routine but also with your overall well-being.

PLANETS

Locate the planets that correspond to the energy you'd like to bring in by sign and by house. Once you've selected a planet you want to activate, use the chart below to select candles, or other objects that resonate with you, in the appropriate colors. Use incense or oils to ramp up the energy and then voila! You are crafting your very own astrological magick.

SUN	Gold, Yellow, Orange	When you need more confidence, or you feel like you are falling away from your true calling. Think about getting a bouquet of sunflowers or dehydrating oranges and making a garland for your window.
MOON	Silver, White, Pale Green	When you need to tap into your feminine side or expand your nurturing qualities. Think about decorating with eucalyptus leaves or putting a silver bauble next to your bed.
MERCURY	Blue, Green, Gray	Great to access for writing, mental clarity, or before a big presentation. Think of doing a watercolor using these colors. Don't do a specific image; just let the colors blend into each other.
VENUS	Pink, Pale Blue, Lavender	Looking to enchant someone or look amazing? Tap into Venus! Think a bath dappled with rose petals or sleeping with a sprig of lavender under your pillow.
MARS	Red, Maroon, Crimson	Great if you're feeling tired, depressed, or low on moxie. Think about applying bright red lipstick or decorating with a bold scarf.

JUPITER	Purple, Indigo, Navy Blue	Great if you're feeling negative and seeing the glass half-empty. Jupiter will increase optimism and luck. Think about investing in some cozy sheets in these calming shades.
SATURN	Black, Gray, Brown	When you need to focus and get a great deal accomplished. Think about setting minerals on a personal altar like tigereye, brown jasper, and tourmaline.
URANUS	Electric Blue, Neon Anything	When you are resisting change, tap into the liberating vibe of Uranus. Think incense smoke and sage smudging.
NEPTUNE	All the Colors of the Ocean	Low on creativity and inspiration? Neptune is incredible for expanding imagination. Think a beach walk if you live by the coast; otherwise seek out any natural body of water and meditate beside it.
PLUTO	Dark Red and Black	When you feel superficial and vapid, bring in Pluto energy for profound depth and intensity. Think connecting with your sexual side using black silk and red roses.

CALLING IN THE ZODIAC

Armed with your newfound knowledge of astrology, you now have several different ways to incorporate the planets into your own special practice of self-care. However, perhaps you'd rather focus on your sign instead. It's equally as potent! Think of using the signs in ritual as more of a "covering all the bases" instead of focusing on a singular planet. Though bonus points for bringing in *both*!

ARIES/FIRST HOUSE

COLORS: Any Shade of Red

OILS: Rosemary, Black Pepper, Clove

ACTIVITY: Light a candle (preferably red), and spend the night creating a vision board. At the end, sit with the vision board and the flame, and let your mind drift to a specific intention. Write it down and repeat if for the next week, month, or however long it takes to engrave itself into your bones and become a part of you!

TAURUS/SECOND HOUSE

COLORS: Yellow, Pale Blue, Pink

OILS: Rose, Magnolia, Citrus

ACTIVITY: Taurus is represented by the bull, which was often used in sacrifices. For this ritual, rest assured, I'm not asking you to kill anything on the altar, but I am asking you to create some quiet space, fill a pretty bowl with roses or magnolias, breathe in the rich scent, and go inward. Ask yourself what you really want, and what you might need to sacrifice to achieve this outcome. Do you want to journal more? Perhaps you need to give up a bit of social media. Are you wanting to focus on yoga? Perhaps you'll need to cut down on binge-watching shows. The answers are inside you, if you are willing to listen.

GEMINI/THIRD HOUSE

COLORS: Violet, Yellow, Blue

OILS: Lavender, Mint, Lemongrass, Clove

ACTIVITY: To channel your Gemini energy, I encourage you to put a few drops of lavender or clove essential oil into a diffuser and inhale the sensual, earthy mist. Set out two opposite materials to focus your intention, one silver and one gold. Touch them. Notice how the light hits each item. How each one brings its own beauty, even though they're different from one another. Meditate on your own dualism, how your very weaknesses might sometimes be your greatest strengths, and vice versa.

CANCER/FOURTH HOUSE

COLORS: Silver, Gray, Green

OILS: Jasmine, Lemon, Rose

ACTIVITY: Fill a bath with jasmine-scented bath salts and soak in a dark bathroom listening to music that makes you feel nostalgic (high school faves, perhaps), or confessional crooners like Adele or Fiona Apple, and let your mind float away on Cancer waves.

LEO/FIFTH HOUSE

COLORS: Orange, Yellow, Red, Gold

OILS: Rosemary, Ginger, Juniper

ACTIVITY: Rosemary is a powerful herb mainly used for protection and cleansing, but also for love and healing. Buy or gather sprigs of it and use it to sweep up your living space, clearing out negative energy. Then hang a bunch over your doorway for added psychic protection.

VIRGO/SIXTH HOUSE

COLORS: Navy Blue, Gray, Purple

OILS: Sage, Patchouli, Lemon

ACTIVITY: Patchouli has a grounding, earthy scent that can be incorporated into a sensual, oh-so-Virgo love and lust ritual. Rub some patchouli soap slowly and intentionally over your body in a hot shower, calling in what it is you crave, be it a lover or any other sensual experience. Close your mind and breathe in the scent, imagining what you want entering the water spray with you.

LIBRA/SEVENTH HOUSE

COLORS: Pink, Yellow

OILS: Chamomile, Rose, Lavender

ACTIVITY: A rose quartz is a pretty pink crystal full of balancing Libra vibes. It's also known to improve sleep quality and channel pleasant dreams. Warm the stone in your hand before bed and focus on any problems you may have. As you drift away, the stone will balance out your anxiety and call in peaceful healing vibes to your bed space.

SCORPIO/EIGHTH HOUSE

COLORS: Maroon Red, Black, Red

OILS: Patchouli, Musk, Ylang-Ylang

ACTIVITY: Ylang-ylang has a witchy, sexy scent with just a touch of darkness, perfect for channeling in those deep Scorpio vibes. Anoint yourself with ylang-ylang–infused oil to ward off seasonal depression or fatigue, and better yet, rub a few drops to the soft skin between your thighs for an extra boost of libido.

SAGITTARIUS/NINTH HOUSE

COLORS: Purple, Dark Blue

OILS: Saffron, Basil, Tea Tree

ACTIVITY: Light a blue candle in a space that feels special to you. Watch the play of the vivid flame against the deep bold color that channels in a sense of wisdom. Reflect on your own inner fire, something Sagittarius knows all about, and meditate on your sagacity, calling in more of the same, into your life.

CAPRICORN/TENTH HOUSE

COLORS: Black, Dark Gray, Blue, Silver

OILS: Ginger, Patchouli, Chamomile

ACTIVITY: Want to call in more Capricorn energy? Treat yourself to a small silver bell. When you need grounding, approach your bell with an intention, then, closing your eyes, ring the bell a few times, letting the resonant sound wash over you, clearing away mental cobwebs.

AQUARIUS/ELEVENTH HOUSE

COLORS: Electric Blue, Neon Yellow

OILS: Pepper, Coriander, Peppermint

ACTIVITY: Grind up coriander seeds and discreetly sprinkle them around people or spaces where you feel out of sorts or don't see eye to eye. Aquarius energy is a little rebellious, but is always a rebel with a cause, so while you might not mind conflict, seeking a resolution with meaningful outcomes is key.

PISCES/TWELFTH HOUSE

COLORS: Sea Green, Indigo

OILS: Sandalwood, Vanilla, Sea Salt

ACTIVITY: Water and salt both have an energetically cleansing power. Put them together to cleanse your aura and allow your Piscean side to vibe. Fill a clear glass with warm water and pour sea salt into it, stirring with a silver spoon. Watch the alchemy at work as the salt gives itself to water, and ponder how you are transforming, retaining your innate self, but also transmuting into new identities.

PEAKS & VALLEYS OF THE MOON

We all get 'em. Okay, so some days or even *months* can seem like nothing but valleys, but this is strictly based on your Sun sign and the daily Moon.

Horrid Saturn transits aside, this is a great tip that I picked up from Debbi Kempton-Smith, taught to her by the brilliant Charles and Vivia Jayne. It's easy and quick, and boy does it work! It's simple, really. First, you'll need access to a daily Moon calendar. (Hot tip! I provide free Moon tables on my site that include void times, too. Head over to rachelstuarthaas.com to find yours!) Now all you need to know is your Sun sign, and remember, *never* ask a favor or make any major decisions during a void Moon. That's it!

You're at your "peak" when the Moon is in your Sun sign and a "valley" when the Moon is opposite your Sun sign. During a peak, it seems like you're more popular, and the Universe just seems to grant your wishes. Take a risk, and schedule a first date or a cosmetic procedure. Put yourself forward for that raise. Start a new project. When the Moon is in your sign a couple of days out of the month, it really is your time to shine. The Universe is listening, so go ahead, *ask*.

Valleys, on the other hand, aren't not necessarily a *bad* couple of days—this isn't Mercury Retrograde, after all—but it is a time where things seem to not always go your way. Other people get perks, their packages get delivered superfast, their food brought out quicker. You're prone to be more tired or have low energy, so take it easy and give yourself a break. The best way to handle a valley is to be of assistance to other people. It'll make you *and* them feel good.

Moon Sign Tables

ARIES PEAK — LIBRA VALLEY
TAURUS PEAK — SCORPIO VALLEY
GEMINI PEAK — SAGITTARIUS VALLEY
CANCER PEAK — CAPRICORN VALLEY
LEO PEAK — AQUARIUS VALLEY
VIRGO PEAK — PISCES VALLEY
LIBRA PEAK — ARIES VALLEY
SCORPIO PEAK — TAURUS VALLEY
SAGITTARIUS PEAK — GEMINI VALLEY
CAPRICORN PEAK — CANCER VALLEY
AQUARIUS PEAK — LEO VALLEY
PISCES PEAK — VIRGO VALLEY

CONCLUSION
LOOK TO THE STARS

The American astrologer Steven Forrest once wrote, "Astrology is just a finger pointing at reality." While many people claim not to be interested in astrology, most people will still admit to knowing their Sun signs. All it takes is an open mind and a willingness to turn inward, to recognize inside each of us is a vast cosmos, made up of stardust, to begin to understand that we are in the Universe, and the Universe is certainly inside us.

I hope that this book has helped you dive deeper into more complex astrological topics like retrogrades, aspects, and asteroids—I also hope that you take your time learning about them and experiencing how they show up for you. Remember that the study of astrology is a marathon, not a sprint.

Pause. Breathe. Savor information and take your time with it all.

You don't want to get overwhelmed and burn out like a supernova! You want to burn slow and steady, like our own Sun, and nourishing your intuition, healing, and clairvoyance.

We all are told about the importance of self-care, to "take care of yourself," as if eating a few greens, putting down the phone at bedtime, and working toward getting those daily ten thousand might seem like good enough. Don't get me wrong—that stuff is good, *real* good! But what about your mental and emotional health? Deliberately engaging in astrology is a way to connect to the universal, planetary bodies that don't move in a time that can fit on a Google calendar, and tell us truths that we won't get in a Zoom call.

This book is an invitation, not just to skim the pages, close the cover, and put it on a bookshelf. It's a tool just like a hammer or a screwdriver. Use it when you have need of it. When something seems broken or not quite right in your life. That's when I want you to pick it back up and reflect: What are the stars telling me? Maybe, just maybe, the Universe is revealing something to you—you just need to take the moment to listen.

Good luck on your journey. I hope you find your adventures in astrology as fulfilling and life-changing as I have, and I hope it's a gift you can share not just with yourself, but with family, lovers, and friends.

The world can be a hard place, but the Universe is always there, and that, my friends, is a comforting thought!

ACKNOWLEDGMENTS

First and foremost thank you to everyone who picked up this book and decided to expand their knowledge of astrology. It's a magical thing to connect with the heavens, and I salute you on your journey.

Thank you so much to my editors Lea Taddonio and Hannah Robinson; there's no way this book would exist without these two! And to the entire team at Simon and Schuster and Tiller Press, Michael Andersen, Lauren Ollerhead, and Molly Pieper: thank you. And many thanks to Nicole Resciniti and Kim Falconer for trusting me with this book. Kim, I am forever in your debt!

Thank you to all the astrologers alive and deceased who've taught me so much and passed on such wise words of wisdom and experience. Jean Norman, Kim Falconer, and the especially the late Debbi Kempton-Smith—my teacher and my mentor. I hope you're dancing up a storm somewhere near Jupiter!

Thank you to all my friends and family who showed up with words of encouragements and support. To Liam and Ivy, who spent their quarantine glued to screens while Mom locked herself in her studio—thank you. And thank you to David, who always makes life extra sparkly.

ABOUT THE AUTHOR

RACHEL STUART-HAAS is a professional artist and astrologer who currently lives in Louisiana. She earned her BFA in design/illustration from the Kansas City Art Institute in Kansas City, Missouri. Since then, Rachel has focused her time and energy on producing one-of-a-kind paintings that portray her intuition toward the obvious and the ethereal. She has been fascinated with astrology since she first laid eyes on Linda Goodman's *Love Signs* in the eighth grade, and after years of working with top astrologers, Rachel launched her own practice, with clients that span the globe. Rachel lives with her husband, David, their children, Ivy and Liam, and her devoted basset hound, Louie.

Visit her online at RachelStuartHaas.com